90

Molecular Biology
Biochemistry and Biophysics
18

Helga Kersten · Walter Kersten

Inhibitors of Nucleic Acid Synthesis

Biophysical and Biochemical Aspects

With 73 Figures

Springer-Verlag New York · Heidelberg · Berlin 1974

Professor Dr. HELGA KERSTEN and Professor Dr. WALTER KERSTEN

Physiologisch-Chemisches Institut
der Universität Erlangen-Nürnberg
D-8520 Erlangen/Fed. Rep. Germany
Wasserturmstr. 5

ISBN 0-387-06825-2 Springer-Verlag New York Heidelberg Berlin
ISBN 3-540-06825-2 Springer-Verlag Berlin Heidelberg New York

Library of Congress Cataloging in Publication Data. Kersten, Helga, 1926. Inhibitors of nucleic acid synthesis. (Molecular biology, biochemistry, and biophysics, v. 18). 1. Nucleic acid synthesis. 2. Antimetabolites. 3. Antibiotics. I. Kersten, Walter, 1926- joint author. II. Title. III. Series. [DNLM: 1. Antibiotics — Pharmacodynamics. 2. Nucleic acids — Antagonists and inhibitors. 3. Nucleic acids — Biosynthesis. W 1 M0195T no. 18 / QU 58 K412i]. QP620.K47. 574.8'732. 74-11265.

Offsetprinting and bookbinding: Brühlsche Universitätsdruckerei, Gießen

Preface

During the last decade physical and chemical methods have improved rapidly - a fact which allowed the mode of action of antibiotics to be studied - and many biochemically-oriented scientists have devoted their research to the following questions:

1. What is the metabolic pathway that is inhibited selectively, and what are the target molecules within a sensitive cell?

2. What are the relationships between the chemical structure of an antibiotic and the physicochemical properties of the sensitive molecule(s)?

3. Why and how far is the action selective?

4. Is it possible to correlate the interaction with the target molecule(s) with the particular biological activities observed?

This monograph deals with those antibiotics which interfere with the biosynthesis of nucleic acids. The idea was to provide an insight into how to investigate the preceding questions experimentally and to solve as yet unresolved problems rather than to give a review of the current state of knowledge. Although the biochemistry of nucleic acid synthesis is known in general, the precise molecular mechanisms by which deoxyribonucleic acid is replicated or transcribed has still to be clarified. For this reason it is not yet possible to describe the molecular mechanisms by which the inhibitors of nucleic acid and protein synthesis exhibit their effects. The fact that the inhibitors of nucleic acid and protein synthesis themselves served as useful tools to obtain an insight into the mechanisms of replication, transcription and translation was one of the most exciting discoveries in this field. Several chapters are devoted to these aspects of antibiotic research.

This discovery that several antibiotics interact with nucleic acids, nucleic acid polymerases or other enzymes, or cell constituents such as ribosomes, allowed biochemists to use these "freaks of nature" as simple model compounds, e.g. for the elucidation of nucleic acid-protein interactions or nucleic acid-carbohydrate interactions. During the last years many physicists have become involved in biological problems; their investigations into conformational changes occurring during the inactivation of biologically active macromolecules will provide us some time in the future with a more precise understanding of how these macromolecules and supramolecular structures exhibit their specific functions. The progress which has been made in the elucidation of the relationships between the chemical structure of an antibiotic and the physicochemical properties of the target molecules, will probably be of help to all those chemists who are concerned with the production of synthetic or semisynthetic compounds for chemotherapeutic purposes. For the question of selectivity of antibiotic action we turn to the virologists, who are concerned with the relationships between the mechanism of antibiotic action and antiviral activity. With increasing knowledge of virus structure and virus replication, an initial picture emerges as to how antibiotics exhibit their selective antiviral versus host activities.

We have taken the risk of stressing biochemical, biophysical, chemical and particular biological aspects. We are aware that we have left many "gaps", e.g. discussion of the biosynthesis of antibiotics and of the mechanisms of antibiotic resistance. After writing this book the authors found their feelings best expressed by EGON FRIEDEL, who said:

"The courage to talk about relationships which are not completely known, to report on facts which are not observed by oneself, to describe processes about which nothing can be said with certainty: in short to say something which could perhaps be wrong, this courage is a prerequisite for all productivity."

Erlangen, October 1974   HELGA KERSTEN · WALTER KERSTEN

Acknowledgements

The authors thank Dr. ALEXIS VON AUFSESS and ANNEMARIE GÖLDEL for most valuable help in preparing the manuscript.

Contents

General Introduction

With the continual discoveries of new antibiotics, the number of compounds that are known to interfere with nucleic acid or protein biosynthesis is steadily increasing. No one knows why different microorganisms produce a variety of substances with complex chemical structures, which are not involved in any metabolic process of the microorganisms and which are extremely toxic. Are they produced: 1. as a chemotherapeutic aid to physicians; 2. to provide physicists with highly sophisticated structures to be resolved; 3. to show chemists by their biosynthetic pathway how to develop new synthetic compounds; or 4. by their specific inhibitory effects, to help biochemists in elucidating as yet unknown aspects of cell metabolism? With respect to the progress in research on DNA, RNA and protein synthesis, e.g. replication, transcription and translation, which has characterized the past 2o years of work in the field of Molecular Biology, the aid provided by antibiotics discussed in the following chapters should be acknowledged.

In this monograph the antibiotics and a few synthetic compounds inhibiting nucleic acid synthesis have been subdivided into four categories and are described in Chapters I - IV:

I. Inhibitors which preferentially affect DNA synthesis

II. Inhibitors which preferentially affect RNA synthesis through interaction with the DNA template

III. Inhibitors which affect RNA synthesis through interaction with RNA polymerases

IV. Inhibitors interfering at the precursor level or with the regulatory processes of nucleic acid synthesis.

For each antibiotic a brief survey on its "origin, chemical and biological properties" is presented, followed by a more detailed description dealing with the interaction of the antibiotic with its target molecule *in vitro*. In this section, "on the molecular mechanism of antibiotic action", the physicochemical methods which can be applied to an investigation of antibiotic nucleic acid or protein interactions are introduced. The techniques available for these binding studies can be divided into two broad categories: 1. nonspectroscopic methods; 2. spectroscopic methods. The nonspectroscopic methods include: equilibrium dialysis, cosedimentation with the target molecule on sucrose gradients or in the analytical ultracentrifuge, gel filtration, electrophoresis, viscosimetry, buoyant density measurements, polarography. Spectroscopic methods include examples of UV and visible absorption, fluorescence spectroscopy, optical rotatory dispersion and circular dichroism, NMR spectroscopy, electron spin resonance, light scattering and X-ray diffraction. A clear comprehensive review of these methods, their principles and their theories has been published by CHIGNELL (1971). Although the methods are discussed with respect to drug-protein interactions, they can in principle be applied to studies of antibiotic nucleic acid or

ribosome interactions. No precise structure of the complexes of antibiotics with nucleic acids or nucleic acid polymerases can be presented. This can, in part, be explained by the fact that the conformation of the macromolecules themselves is not precisely known.
Eleven years after the discovery of the "DNA-actinomycin complex"
(KIRK, 1960; KERSTEN et al., 1960; KAWAMATA and IMANISHI, 1960)[1], the
structure of a complex of actinomycin with a synthetic double-stranded
oligo-nucleotide, was elucidated by X-ray crystallography (SOBELL et
al., 1971).

The section which follows the discussion on the molecular mechanism
of action deals with the inhibitory effects on macromolecular synthesis or metabolism in prokaryotic or eukaryotic cells. Questions
are raised as to whether the interaction with a target molecule can
be selective in such a complex system as the cell, and how far other
cell constituents might compete with the inhibitor for the binding
site.

Several of the antibiotics have "particular biological properties"
which are described in a separate section. The problem as to how
far the particular biological effects can be correlated to the proposed molecular mechanism of action, still remains to be solved in
many cases.

The last category, IV, involves nucleoside analogues or amino acid
analogues which are more or less specific inhibitors of nucleic acid
synthesis and can interrupt other metabolic pathways and protein biosynthesis. Some quinone antibiotics and mitomycin derivatives specifically interfere with the synthesis of amino acids, which control
RNA synthesis via the RC or rel gene in bacteria.

The molecular mechanisms by which inhibitors of nucleic acid synthesis interfere with cell metabolism have been discussed more
comprehensively by WARING (1972).

---

[1]References, see Chapter II/1

I. Inhibitors of DNA Synthesis

The mechanism of DNA replication in prokaryotic and even more in
eukaryotic cells is a complex process in which apparently more than
one DNA polymerizing enzyme system is involved.

The first DNA polymerizing enzyme was isolated and characterized in
Kornberg's laboratory (for review see KORNBERG, 1969). The DNA
polymerase I - from E.coli - has a molecular weight of 1o9,ooo,
consists of a single polypeptide chain, binds to single-stranded
linear or circular DNA, to nicked double-stranded but not to double-
stranded circular and double-stranded linear DNA. The enzyme contains
one binding site to which each of the four deoxyribonucleotide
triphosphates can bind. The enzyme catalyzes several reactions: DNA
polymerization from 5' phosphate to 3' OH of deoxyribose, pyrophosphate exchange and pyrophosphorolysis; the enzyme exhibits exonuclease activity and is involved in the excision repair mechanism.

Besides the E. coli DNA polymerase I, several other in vitro systems
capable of DNA synthesis have been described (GANESAN, 1968; SMITH
et al., 197o; KNIPPERS and STRÄTLING, 197o; KORNBERG, T. and GEFTER,
197o; OKAZAKI et al., 197o; KNIPPERS, 197o; MOSES and RICHARDSON,
197o; GANESAN, 1971). The studies indicate that DNA synthesis in
vitro is not necessarily related to DNA replication (CAIRNS, 1972).

DNA polymerase II and DNA polymerase III have been purified from
E.coli mutants lacking DNA polymerase I. Like the DNA polymerase I,
both enzymes catalyze in vitro a repair type of reaction (RICHARDSON,
1972).

With the possible exception of Ganesan's system (GANESAN, 1971), DNA
synthesis differs from replication in all the open in vitro systems
reported: For example in sensitivity to specific inhibitors, in the
formation and joining of Okazaki pieces (OKAZAKI et al., 1968), in
temperature sensitivity, in kinetics of synthesis, in the rate of
synthesis, in the rate of chain elongation, etc. SCHALLER et al.
(1972) describe an in vitro system for DNA replication which uses
a highly concentrated lysate of E.coli deficient in DNA polymerase I.
This lysate differs from other systems in that it contains all the
macromolecular components of the bacterial cell at very high but
ill-defined concentrations. The DNA synthesis observed resembles DNA
replication in almost all of the respects investigated.

At present it is assumed that in bacteria one strand of the DNA is
replicated continuously, the other strand discontinuously. OKAZAKI
et al. (1973) reported that during discontinuous replications
attachment of short RNA chains to nascent DNA fragments occurs. The
size of the RNA chains is estimated to be about 5o-1oo ribonucleotides. ALBERTS et al. (1973) presented evidence that the replication
of T_4 DNA involves six proteins coded for by the T_4 phage genes. The
proteins were isolated and characterized from T_4-infected E.coli. An
in vitro DNA-synthesis system was developed which shows the requirements for these gene products. One of the proteins is thought

to be responsible for exposing the template strands and aligning the
bases in a conformation which is optimal for rapid polymerase action.
Some additional proteins required for T$_4$ DNA replication could have
purely structural roles; others are probably needed for the frequent
de novo chain initiations.

Investigations on replication in eukaryotes still reveal a confusing
picture. Evidence has been presented which indicates that nearly all
replicated DNA is associated with membrane material - which suggests
that DNA replication sites are located at the nuclear membrane. In
contrast, electron-microscopic autoradiographic experiments indicate
that DNA replication sites are distributed throughout the nucleus.
It was therefore concluded that initiation of DNA replication takes
place on the nuclear membrane, while replication itself might occur
throughout the nucleus.

The inhibitors of DNA synthesis discussed in this chapter represent
a variety of chemically distinct compounds, including the alkylating
agent aziridine, mitomycin, the quinone-containing streptonigrin and
different types of polypeptides: 1. the bleomycin/phleomycin group,
2. the spermidine-containing edeines and 3. neocarcinostatin. Most
of the inhibitors of DNA synthesis discussed in this chapter inhibit
DNA replication in prokaryotic as well as in eukaryotic cells. With
the exception of the peptide antibiotics neocarcinostatin and edeine,
they affect the DNA polymerase I reaction *in vitro* because they bind
to DNA. Recently mitomycin and nalidixic acid have been shown to inhibit DNA synthesis with DNA polymerizing system described by SCHALLER
et al. (1972). No doubt future studies with the different inhibitors
of DNA replication will help us to understand the mechanism of DNA
replication of phages and viruses and DNA replication in bacteria and
higher organisms. The peptide antibiotics will be extremely useful in
the elucidation of the binding forces involved in the interaction of
nucleic acids and proteins.

Moreover specific inhibitors for lymphocyte DNA synthesis which can
be isolated from the lymphoid system and might occur also in melano-
cytes, granulocytes, kidney cells and cells of the lens can become
useful therapeutic agents (HOUCK et al., 1971).

A. Mitomycin

1. Origin, Biological and Chemical Properties

The mitomycins were discovered by HATA et al. (1956) in *Streptomyces
caespitosus* and isolated by WAKAKI et al. (1958) in the form of deep
violet crystals. Porfiromycin, produced by *Streptomyces ardus*
(DE BOER et al., 1961) also belongs to this group. The two unidentified antibiotics G 253B and G 253C from *Streptomyces reticuli* var.
shimofasuensis described by HATA et al. (1966) resemble the mitomycins
in molecular weight, absorption spectra and elemental analysis.

Depending on their concentrations the mitomycins exhibit a variety
of inhibitory effects. At very low, non-inhibitory concentrations
they suppress the ability of recipient *Diplococcus pneumoniae* cells
to integrate transforming DNA. Noncidal concentrations inhibit cell
division of several bacterial species and of mammalian cells. Higher

mitomycin concentrations result in bacteriostasis and in rapid killing. The mitomycins have been shown to be active against some viruses, to exhibit mutagenicity, to induce lysogenic phages and to cause the production of colicins. (See review by SZYBALSKI and IYER, 1967.) Mitomycin C suppresses the transformation of normal plant cells by the crown-gall organism *Agrobacterium tumefaciens* without macroscopically observable phytotoxic effects when applied to *Kalanchoe daigremontiana in vivo* (GRIBNAU and VELDSTRA, 1969). Among several groups of carcinostatic substances, the mitomycins constitute a most powerful group (see review by CARTER, 1968 and MOORE et al., 1968).

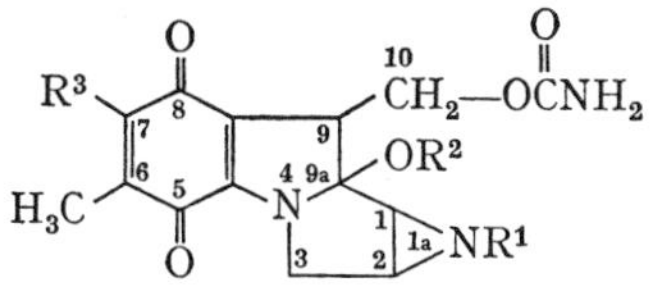

	R^1	R^2	R^3
Mitomycin A	H	CH_3	H_3CO
N-Methyl-mitomycin A	CH_3	CH_3	H_3CO
Mitomycin B	CH_3	H	H_3CO
Mitomycin C	H	CH_3	H_2N
Porfiromycin (N-methyl-mitomycin C)	CH_3	CH_3	H_2N
7-Hydroxy-porfiromycin	CH_3	CH_3	HO

Fig. 1. Structure of mitomycins and porfiromycins. (According to WEBB, J.S., et al. (1962). From: J. Amer. Chem. Soc. 84, 3185 (Antibiotics I, Eds.: D. GOTTLIEB and P.D. SHAW. Berlin-Heidelberg-New York: Springer, 1967, p. 211))

The chemical structures of the mitomycins, shown in Fig. 1, were elucidated by WEBB et al. (1962). All mitomycins, including porfiromycin, have structure I in common, differing only in minor substituents.

The mitomycins represent the first naturally occurring compounds containing an aziridine ring, a pyrrolo (1,2-a) indole ring system, an amino- or methoxybenzoquinone and pyrrolozidine residue. All members of the mitomycins have the three typical carcinostatic groups, i.e. aziridine, quinone and urethane. In contrast to the usual synthetic ethyleneimine rings, the aziridine ring is stable and inactive in the native form of the mitomycin molecule (SCHWARTZ et al., 1963). Conversion to the active state occurs by a reductive process (SCHWARTZ et al., 1963; PATRICK et al., 1964; SZYBALSKI and ARNESON, 1965). The alkylating properties of the mitomycins are strictly dependent on the aziridine ring system.

2. On the Molecular Mechanism of Action

As stated previously, the mitomycins contain three main functional groups. The antibiotics may therefore interact differently with certain cell components. Possibly the nutritional environment of the cells determines which functional group of the molecule becomes active. Thus some selectivity for one or another inhibitory effect should be considered. As a consequence, the mechanism of action need not be unique and might depend on 1. the cell type, 2. the physiological state of the cells and 3. their environment.

a) Interaction with DNA *in Vitro*

In vitro the mitomycins react with purified DNA in the presence of reducing agents or NADPH and a quinone reductase. DNA treated in this way renatures spontaneously on heat denaturation and rapid chilling, retaining its native-like hypochromicity, transforming activity and buoyant density in cesium chloride - or cesium sulphate - gradients. IYER and SZYBALSKI (1963, 1964) concluded that activated forms of mitomycins alkylate DNA, thereby forming cross-links between the two strands of the DNA molecule. The experimental evidence for this conclusion is shown in Fig. 2.

Among several bacterial DNAs tested, the highest susceptibility to mitomycin cross-linking was observed with *Sarcina lutea* DNA (7o% G+C) and the lowest susceptibility with DNA from *Cytophaga johnsonii* (32% G+C).

Purified *Bacillus subtilis* DNA binds one mitomycin molecule per 2,5oo nucleotide pairs, i.e. per molecular weight on 1.5×10^6 daltons. Only one out of five to ten antibiotic molecules participated in the cross-links, and the others reacted with one strand only. The mitomycins were thought to interact with DNA in the same manner as the difunctional alkylating agents studied by LAWLEY and BROOKES (1963, 1967).

WEISSBACH and LISIO (1965) showed that up to one antibiotic molecule can be bound per 5oo nucleotide residues of DNA from *E. coli*. Ribonucleic acids and ribosomes are also alkylated but to a lesser degree than DNA. Bovine plasma albumin, starch and glycogen can be alkylated by mitomycins under comparable conditions, but to a considerably lesser degree than nucleic acids. The possible site of alkylation of nucleic acids by mitomycins was studied by LIPSETT and WEISSBACH (1965). Experiments with synthetic polyribonucleotides showed that alkylation with ^{3}H-mitomycin or ^{14}C-porfiromycin proceeds at least four times as easily on guanine as on the other common bases. Alkylation of tRNA with porfiromycin and subsequent hydrolysis yields demonstrable quantities of both monoguanyl- and diguanyl-porfiromycin. It was suggested that the diguanyl product probably arises from interstrand cross-linking. However analogous experiments with DNA have not yet been reported.

Interstrand cross-linking of DNA by difunctional alkylating agents has been extensively studied. LAWLEY and BROOKES (1967) found that double-stranded but not single-stranded DNA reacted *in vitro* with difunctional alkylating agents such as S-mustard. Upon hydrolysis, di-(7-guanyl-) derivatives resulted. Therefore the product formed must originate from interstrand cross-links. In addition to cross-linking, part of the difunctional alkylating agents alkylate N 7 of

---

►

Fig. 2. Microdensitometric tracings of photographs taken after 4o hr of Cs$_2$SO$_4$ equilibrium density gradient centrifugation (31,41o rpm, 25 C) of DNA extracted from *B. subtilis* cells never exposed to mitomycin (*A*, *B*) or grown for 15 minutes in the presence of 12.5 µg mito-

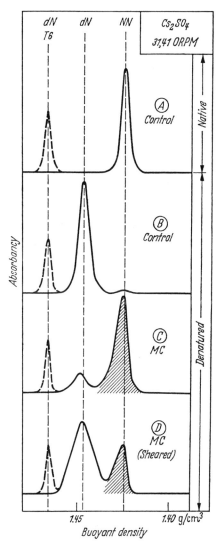

mycin C/ml of nutrient broth (C, D). The DNA was either native (A)
or denatured (B, C, D) by 6 minutes exposure to 100°C in 0.02 M Na$_3$·
citrate buffer at pH = 7.8 followed by rapid chilling at 0°C. The
DNA used in experiment D was mechanically sheared prior to denaturation by forcing a solution of 15 µg DNA/ml three times through a
25-gage needle (I.D. = 0.254 mm) with a spring-loaded, constant-rate
syringe CR 700 (Hamilton Co., Whittier, Calif.). The shaded areas
correspond roughly to mitomycin cross-linked DNA. The broken lines
(DN $T6$) represent denatured T6 coliphage DNA added as a density marker. The buoyant densities of native (A) and denatured (B) $B.$ $subtilis$
DNA correspond to 1.424 (NN) and 1.446 (dN) g/cm^3, respectively. The
proportion of denatured, non-cross-linked DNA, determined by comparing the areas under the peaks, increases upon shearing from 23% (C)
to 73% (D). (According to SZYBALSKI and IYER, 1964 a. From: Antibiotics I. Eds.: D. GOTTLIEB and P.D. SHAW. Berlin-Heidelberg-New York:
Springer 1967, p. 221)

guanine as do monofunctional alkylating agents. Small amounts of 1-alkyladenine and 3-alkyladenine have also been observed.

Because mitomycin shows preference in binding for G+C-rich DNA, it was suggested that diguanyl-mitomycin participated in interstrand cross-linking. Space filling models of DNA and of mitomycin showed that it is extremely difficult to fit the mitomycin molecule as a cross-link between the complementary strands of DNA without assuming a large distortion of the double helix. The best fit with the models was obtained by postulating links between two O6 groups of the nearest guanines on the opposing strands (SZYBALSKI and IYER, 1967).

A new assay to measure 7-alkylation of guanine residues in DNA was described by TOMASZ (1970). This assay is based upon the finding that methylation of the N7 position of guanine derivatives renders the C8 hydrogen extremely labile. This lability is manifested by a rapid deuterium or tritium exchange of the C8 hydrogen with the solvent under physiological conditions. The method was used to analyze the interaction of DNA with three drugs: the bifunctional nitrogen mustard: methyl-(bis-β-chlorethyl)-amine, the synthetic mutagen triethylenemelamine and the antibiotic mitomycin C. The nitrogen mustard showed a stoichiometric correspondence between the amount of guanine alkylated and the amount of tritium released into the medium. Triethylenemelamine caused substantial release of tritium. There was however no detectable tritium release with mitomycin C under the optimal conditions for covalent binding of the drug to DNA *in vitro*. These findings indicate that, contrary to earlier suggestions, mitomycin differs from the difunctional alkylating agents in its interaction with DNA.

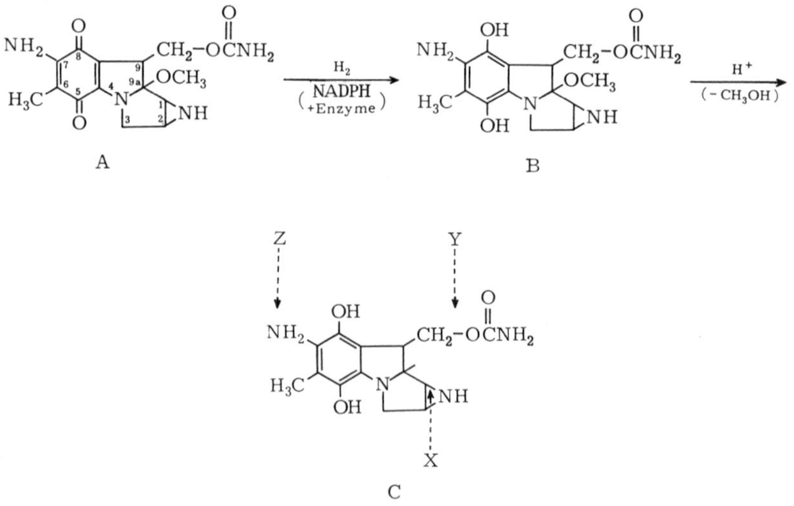

Fig. 3. Structure of mitomycin (*A*), the primary product after chemical or enzymatic reduction (*B*), postulated structure after secondary rearrangement (*C*), *X* and *Y* and *Z* are the possible reaction sites of the difunctional alkylating product. (According to SZYBALSKI, W., and IYER, V.N., 1967. From: Inhibitors, Tools in Cell Research, Mosbach Colloquium 1969. Berlin-Heidelberg-New York: Springer 1969, p. 15)

The active alkylating species of reduced mitomycin C is not known
with certainty. The postulated mechanism of activation is shown in
Fig. 3 (for details see SZYBALSKI and IYER, 1967). The reduction of
the quinone moiety of mitomycin is followed by spontaneous elimination
of the tertiary methoxy group and formation of an aromatic indole system with two active centers. The reactive sites are probably the open-
ed aziridine ring and the CH_2 group of the methylurethane side chain.
The substituent in position 7 of the quinone ring might also be involved in the interaction with DNA.

An electronic mechanism of cross-linkage formation was discussed extensively by MURAKAMI (1966). The mitomycins are completely inert in
their native form; reduction results in the extension of a π-electron
system in their molecular structures. The sensitivity of the aziridine
ring during the activation reaction is considered to be caused by an
electronic effect on that ring. This theory explains the high reactivity of the postulated ionized radical system of the activated molecule, corresponding to the intermediary semiquinoid form. This form of
the mitomycin molecule exhibits, to quote the authors, "beautiful
harmony with the known electronic features of the coupled bases in
DNA and hence interacts with them as a bi- or even polyfunctional
agent". The ability of the mitomycin molecule to form the cross-linkage on the double strands of DNA is interpreted on the basis of a
radical mechanism including the electronic transformations mentioned
above.

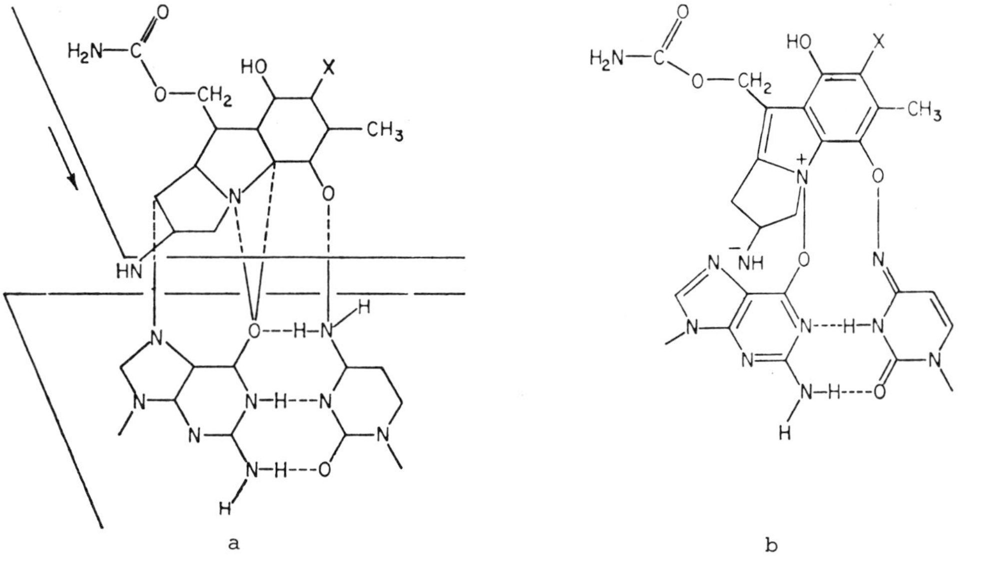

a   b

Fig. 4. *a*) Electronic interactions between the cytosine and guanine
pair in DNA and the active form of mitomycin, approaching with the
suitable relative orientation. The bi- or polyfunctional feature of
the activated mitomycin is remarkable.
b) One example of the cross-linking of the mitomycin molecule with
the coupled bases in DNA. As a result of these linkings, the structures of guanine and cytosine are changed to their tautomeric isomers.
(According to MURAKAMI, 1966. From J. Theoret. Biol. 1o, 236 (1966))

This postulated scheme of cross-linkage formation results in a tautomeric change in the purine and pyrimidine bases in DNA. The greater
tendency to such isomerization exhibited by cytosine and guanine as
compared with thymine and adenine is thought to provide a basis for
explaining the experimentally observed preferential contribution of
cytosine and guanine to the cross-linking reactions. This postulated
mechanism of formation of cross-links does not involve covalent binding of mitomycin to the N7 of guanine. Thus, this model can help to
explain the more recent findings of TOMASZ (197o) that mitomycins do
not alkylate the N7 position of the guanine residues in DNA as do the
other difunctional alkylating agents.

b) Interaction with DNA within the Cell

The renaturation following rapid cooling of heat-denatured DNA extracted from mitomycin-treated microorganisms was demonstrated by
IYER and SZYBALSKI (1964), indicating that mitomycins cause cross-
links *in vivo* as well as *in vitro*. Exposure of *B.subtilis* cells to
N-methyl mitomycin-^{14}C results in a binding of the antibiotic to intracellular DNA. Changes in the physicochemical properties of DNA
isolated from mitomycin-treated cells were reported by WHITE and WHITE
(1965). IYER and SZYBALSKI studied the transforming activity of DNA
isolated from mitomycin-treated *B. subtilis*. Even when the cross-linking
was extensive, as much as 2o% of the original transforming activity
was still retained by the molecules. Furthermore, the biological activity of the linked DNA, unlike the activity of normal DNA, was not
critically and abruptly destroyed over a narrow temperature range.
The remarkable heat stability of the DNA isolated from mitomycin-
treated cells indicates that this residual transforming activity is
primarily associated with the cross-linked DNA molecules. Prolonged
exposure to mitomycin can, however, result in a loss of all detectable
transforming activity of the DNA (TERAWAKI and GREENBERG, 1966). The
transforming activity of DNA from streptococci was also found to be
depressed upon prolonged treatment of the cells with mitomycin C (TE-
RESHIN, 1969).

It is now well established that damage caused in DNA by X-irradiation,
UV-irradiation, by mono- or difunctional alkylating agents and by
mitomycin can be repaired. For reviews, see SETLOW (1968), HOWARD-
FLANDERS (1968), HANAWALT (1968), STRAUSS (1968) and RAUTH (1966).
X-irradiation results in chemical changes of the pyrimidines. UV-
irradiation leads to the formation of covalently linked pyrimidine
dimers, mainly thymine dimers (BEUKERS and BERENDS, 196o, and WACKER,
1963). Damage by mono- and difunctional alkylating agents is directed
mainly to the N7 of guanine. Other bases or even the phosphate groups
may be alkylated (FREESE, 1969).

c) Effect on DNA Metabolism

The selective inhibitory effect of mitomycin C on the synthesis of
DNA in prokaryotes and eukaryotes has been described by several investigators (SHIBA et al., 1958; REICH and FRANKLIN, 1961; REINMER
and YOSHIDA, 1968; ALBACH and SHAFFER, 1967). The inhibitory effect
of mitomycin on DNA synthesis is suggested to be a consequence of the
cross-linking effect of mitomycin which prevents the separation of
the two strands of DNA during the replication process (SZYBALSKI and
IYER, 1964).

Mitomycins induce DNA breakdown so that concurrent depolymerization and synthesis can take place. A depolymerization of DNA in mitomycin-treated microorganisms or mammalian cells has been observed (KERSTEN, 1962 a, b; KERSTEN and THEMANN, 1962; SHATKIN et al., 1962; MATSUMOTO et al., 1966). This depolymerization depends on the concentration of magnesium ions in the culture medium (KERSTEN and KERSTEN, 1963) and leads to the formation of mononucleotides and free bases (REICH and FRANKLIN, 1961; SHIIO et al., 1962). Little or no mitomycin-induced DNA breakdown was noticed in a number of other mammalian or bacterial cultures (IYER and SZYBALSKI, 1963; MAGEE and MILLER, 1962; BOYCE and HOWARD-FLANDERS, 1964; ALBACH and SHAFFER, 1967). The depolymerization of DNA in mitomycin-exposed cells probably depends on the capacity of these cells to repair_DNA, i.e. on the activity and amount of re-pair enzymes. *E. coli* uvr⁻ mutants, which are hypersensitive to mito-mycins, do not show considerable breakdown of DNA at lethal concentra-tions whereas the DNA of mitomycin-resistant uvr⁺ mutants becomes par-tially degraded. The excision-repair which is accompanied by degrada-tion of DNA is enhanced by chloramphenicol, indicating that the ex-cision steps, but not the subsequent steps of the repair process, i.e. "DNA synthesis", can occur in the absence of protein synthesis (PA-PIRMEISTER and DAVISON, 1965). This result could explain the findings of KERSTEN and KERSTEN (1963) and CONSTANTOPOULOS and TCHEN (1964) that the mitomycin-induced breakdown of DNA in the resistant *E. coli* is markedly enhanced by chloramphenicol.

An increase in the activities of DNA-degrading enzymes in mitomycin-treated microorganisms was observed by NAKATA et al. (1962), KERSTEN and KERSTEN (1963), KERSTEN et al. (1964), LEOPOLD et al. (1965), NIITANI et al. (1964), STUDZINSKI and COHEN (1966), and STUDZINSKI et al. (1966). The increased activities of DNAases in mitomycin-treated HeLa cells may be related to repair processes.

d) Effect on RNA Metabolism

The mitomycins not only interfere with DNA and DNA metabolism, but also affect RNA metabolism (SMITH-KIELLAND, 1966 a, b).

On prolonged treatment of *E. coli* with mitomycin C, ribosomes, ribo-somal RNA and sRNA become degraded (KERSTEN and KERSTEN, 1963; KERSTEN et al., 1964; LEOPOLD et al., 1965). Besides a degradation of ribo-somes, SUZUKI and KILGORE (1964) observed a reduced rate of ribosomal RNA synthesis when *E. coli* was treated with mitomycin. Simultaneous synthesis and degradation, i.e. a rapid turnover of RNA, occur in the presence of mitomycin, as was described by KATO et al. (1970) under conditions in which DNA remained unaffected during the treatment. These authors present evidence to suggest that RNAase I does not par-ticipate in this degradation of ribosomes. In a mutant strain of *Al-caligenes faecalis* not containing RNAase I, RNA and ribosomes are also degraded (NATORI et al., 1967). Thus it seems improbable that RNAase I participates in the degradation of RNA, although increased "overall RNAase activity" in cell extracts from mitomycin-treated *E. coli* was found (KERSTEN and KERSTEN, 1969).

The inhibitory effect of mitomycins on the synthesis of RNA is ex-plained by KERSTEN and KERSTEN (1969) as involving the quinone ring, since mitomycin derivatives not containing the alkylating aziridine group and synthetic quinones have been found to inhibit, immediately

84020

and reversibly, the synthesis of proteins and RNA in microorganisms. The mechanism by which certain quinone antibiotics interfere with protein synthesis is shown to involve an inhibition of the amino acylation of certain tRNAs. As a consequence the phenomenon of stringent response is evoked, e.g. formation of ppGpp and inhibition of RNA synthesis. (For further details see Chapter IV.)

3. Particular Activities in Biological Systems

a) Synthesis of Enzymes

Mitomycins inhibit the synthesis of induced enzymes (CHEER and TCHEN, 1962, 1963; CUMMINGS, 1965; COLES and GROSS, 1965; KIT et al., 1963; SHIBA et al., 1958; TAKAGI, 1963). This inhibitory effect on the induced synthesis of, for example penicillinase, occurs at concentrations, and at a time, at which RNA synthesis is arrested following the inhibition of the synthesis of DNA. According to BASU et al. (1965) it seems likely that the inhibition of induced β-galactosidase synthesis by mitomycin C is a case of enhanced catabolite repression. Since catabolite repression of β-galactosidase synthesis appears to be caused by an inhibition of the transcription of the *lac* gene, the effect of mitomycin on the rate of β-galactosidase messenger RNA synthesis was followed indirectly by measuring enzyme synthesis. Mitomycin caused appreciable reduction of the transcription of the *lac* gene in the presence of glycerol, but stimulated the transcription nearly twofold in its absence.

Mitomycin activates thymidine kinase in two-day-old human embryonic lung cells in culture. Additional treatment with an inhibitor of protein synthesis, puromycin, in combination with mitomycin, prevented the elevation of thymidine kinase activity. This indicates that mitomycin somehow induces the synthesis of this enzyme (ROSS and SOLYMOSI, 1967).

LERMAN and BENYUMOVICH (1965) studied the effect of mitomycin C on overall protein synthesis in human neoplastic cell lines. These authors found a rapid inhibition of protein synthesis following the addition of mitomycin, which was comparable with that observed upon addition of puromycin. Probably the function of the quinonering predominates in tumortissue and is responsible for the inhibition of the proteinsynthesis. (For further details see Chapter IV.)

b) Mutagenicity

The mitomycins are mutagenic for bacteria (SZYBALSKI, 1958; IIJIMA and HAGAWARA, 1960; TSUKAMURA and TSUKAMURA, 1962). The mutagenic effect of the mitomycins can easily be explained by a direct interaction with cellular DNA. The nature of the damage to the DNA that causes the mutagenic effect remains to be elucidated.

c) Chromosome Breakage

Mitomycins frequently produce cytologically-detectable "translocation-like" cross configurations in dividing leukocytes (COHEN and SHAW, 1964). Similar results were obtained by NOWELL (1964) following treatment of leukocyte cultures with mitomycin. From the cytological data, SHAW and COHEN (1965) suggested that a process analogous to the somatic crossing-over and also to reciprocal translocations between nonhomologous autosomes can be induced by mitomycin. A cytogenic effect of mitomycin C on chromosomes in cultured human cells was described by SINKUS (1969). Mitomycins cause mitotic inhibition and chromosome damage involving chromosomes 1, 9 and 16 in the paracentromeric region of the long arms. GERMAN and LA ROCK (1969) also described the effects of mitomycins on chromosomes and maintain that these antibiotics are potential recombinogens in mammalian cell genetics. Mitomycins produce localized chromosome breaks also in plants, e.g. in the heterochromatin segments of *Vicia* root cells (ARORA et al., 1969).

The stimulation of chromosomal exchanges, crossing-over and the increased rate of genetic recombination observed in mitomycin-treated organisms, favor the hypothesis that the biological effects of mitomycins on chromosomes involve DNA-specific enzymes and are probably not caused by random alkylation of the bases or cross-links in the DNA molecule. Fragmentation of chromosomes is probably the result of enzymatic DNA breakdown.

d) Viruses, Phages and Episomal DNA

Mitomycins induce lysogenic phages (KORN and WEISSBACH, 1962; LEIN et al., 1962; LEVINE, 1961; OTSUJI, 1961, 1962). *E. coli* K_{12} (λ) lyse within 9o minutes after treatment with mitomycin with optimal production of phages.

It is presently known that various phage-inducing agents such as UV cause damage in DNA. The development of phages stops when the damaged DNA is repaired by the excision-repair mechanism. TAKENO et al. (1968) reported that the mitomycin-induced phage formation and lysis of cells can be suppressed when the host cells, K_{12} (λ), are exposed to light. The mechanism of this photosuppression involves the photoreactivating enzyme for DNA repair.

B. subtilis 168 and related strains have been reported to carry defective phages, e.g. PBSX (SEAMAN et al., 1964); phage µ (IONESCO et al. 1964); and "phage-like particle" (STICKLER et al., 1965). Although the particles kill certain sensitive *B. subtilis* strains and seem to share the properties of bacteriocins and temperate phages, no bacterial strains have been found to support their growth. Phage-like particles can be induced in the absence of DNA synthesis, which suggests that most of the DNA in the particles originates from the bacterial chromosome (SEAMAN et al., 1964). OKAMOTO et al. (1968 a, b) studied in more detail the phage-like particles, obtained following mitomycin treatment of *B. subtilis*. The particles were purified by sucrose gradient and cesium chloride density gradient centrifugations. The particles adsorb to cells of sensitive strains, resulting in a killing pattern resembling a single hit process. There is no detectable injection of DNA from the particles into the bacteria. The DNA within the phage heads has a high degree of size homogeneity with a sedimentation coefficient, $S^0 2o$, w, of 22. DNA-DNA hybridization and transformation experiments indicate that most if not all of the DNA in the phage-like particles is of bacterial origin.

These results clearly show that, after induction with mitomycin, bacterial DNA is degraded, converted into fragments of homogeneous size and wrapped into coat proteins. As much as 5o% of the pre-labelled bacterial DNA can be incorporated into the phage-like particles (OKAMOTO et al., 1968 b). The fact that chloramphenicol inhibits the appearance of 22S DNA indicates that some protein is necessary for the specific breakdown of DNA. Evidence is presented to show that the synthesis of this specific protein starts between 3o and 4o minutes after the addition of mitomycin. The formation of the phage-like particles can be explained by the induction by mitomycin of a specific DNA-degrading enzyme. As early as 1957, FREDERICQ drew attention to the similarities between colicinogenic factors and lysogenic phages, and suggested that the process of colicin induction involved a vegetative multiplication of colicinogenic factors. DE WITT and HELSINKI (1965) presented experimental evidence to show that the factor col E_1 from *E. coli* is a satellite DNA which increases quantitatively under conditions of induction by mitomycin C. Furthermore, there was approximate proportionality between the magnitudes of the increase in col E_1 and the colicin production. The striking analogies between the induction of replication of col E_1 and the induced multiplication of bacteriophages favor the view that the increased production of colicin E_1 is at least partly due to the increase in copies of the genetic determinants for col E_1. Colicinogenic factors are autonomous DNA. The replication of this DNA might start immediately upon inhibition of host-cell DNA synthesis by mitomycin C. A different sensitivity of episomal DNA versus host DNA to mitomycin has been found in *E. coli* containing the sex factor F^+. Mitomycin inhibits the synthesis of both host DNA and episomal DNA. The host-cell DNA becomes degraded whereas F^+ DNA is resistant to degradation (DRISKELL-ZAMENHOF and ADELBERG, 1963). Several virus/host systems are described in which the sensitivity of virus DNA and host DNA differs (for summary see review by SZYBALSKI and IYER, 1967). Thus the inducing effect of mitomycin on the release of lysogenic phages, the production of "phage-like particles" and the formation of colicins are most probably related to the specific inhibitory effect of mitomycin on the synthesis of host DNA and the induction of specific host DNA-degrading enzymes. Favoring this view are the observations of LINDQVIST and SINSHEIMER (1966), who found that host DNA synthesis in mitomycin-treated repair-deficient hcr⁻ *E. coli* mutants was selectively inhibited without affecting the synthesis of viral DNA (Φ X 174).

Some RNA viruses containing single-stranded RNA can be produced in the presence of mitomycin (COOPER and ZINDER, 1962; REICH and FRANKLIN, 1961; KNOLLE and KAUDEWITZ, 1964). ROTT et al. (1965) showed that mitomycin inhibits the growth of the double-stranded RNA of fowl plague myxovirus, and VIGIER and GOLDE (1964 a, b) found the RNA Rous sarcoma virus to be sensitive to mitomycin.

The inhibitory action of mitomycins on RNA viruses either depends on a direct effect of mitomycins on viral RNA synthesis, or may be caused by a rapid degradation as a consequence of RNAase induction. If the viral RNA is replicated by intermediate formation of RNA-like DNA by an RNA-dependent DNA polymerase (TEMIN and MIZUTANI, 197o), mitomycin might also affect viral RNA synthesis by interacting with the intermediately-formed DNA.

e) Mitosis

Certain chemicals have different toxicities with respect to various stages of the cell division cycle. This has been observed in several types of mammalian cells in culture for vinblastine (BRUCHOVSKY et al., 1965), hydroxyurea (SINCLAIR, 1965), and 5-bromodeoxyuridine (KIM et al., 1967). During the division cycle, HeLa cells were found to be most susceptible to mitomycin in the G_1 phase whereas their sensitivity to actinomycin D was most pronounced in the S phase. Very little cross-linking was found in DNA isolated from mammalian cell lines (DJORDJEVIC and KIM, 1968).

PARKIN and CHIGA (1966) investigated the effect of mitomycin on hepatic regeneration. They found that under certain experimental conditions mitomycin allows DNA synthesis in rat liver after partial hepatectomy, but blocks mitosis completely. The inhibition of mitosis by mitomycin C is thus not correlated with the inhibitory effect on DNA synthesis.

The precise mechanism of mitosis is not yet known. Therefore the mechanism by which several antibiotics interfere with mitosis cannot be explained. It has been postulated that the cell membrane is the key structural component involved in the initiation of cell division. Some authors speculate that mitomycin may alkylate membranes or other cytoplasmic components involved in mitosis (DJORDJEVIC and KIM, 1968; GRULA et al., 1968; SZYBALSKI, 1964). An interesting observation in this respect was made by STEIN and ROTHSTEIN (1968): mitomycin blocks mitosis in cultured frog lenses, even when the drug is added to the system after peak DNA synthesis. The authors suggest that mitotic inhibition produced by the antibiotic may be due to an effect on the synthesis of RNA.

f) Immune Response

Many agents that are clinically useful in the treatment of cancer possess immunosuppressive activity. The action of mitomycin on DNA synthesis, and thus on cellular proliferation, would account for its bone marrow depressive activity and would suggest immunosuppressive action. It has been shown conclusively by SAKAUCHI and DE WITT (1967) that mitomycin inhibits antibody production when given in minute doses at the time of antigenic stimulation, but not when given at the height of the immune response; this confirms the absence of any direct action of mitomycin on protein production. LEMMEL and GOOD (1969) studied the tolerance of cell-mediated immune responses after *in vitro* treatment of competent cells with mitomycin C. Their findings indicate that *in vitro* treatment of lymphoid cells with mitomycin C prevents graft-versus-host reactivity, as can be shown by injection into appropriate recipients. The treated cells, however, seem to survive *in vivo* and to maintain cellular reactivity towards such party antigens as are detected by appropriate transfer studies. The findings indicate that mitomycin can induce a temporary blockage of certain activities of the treated cell population, permitting recovery of at least some of the injected cells. This recovery leads to the development by these cells of specific tolerance to the antigens of the recipients. In good accord with the immunosuppressive action of mitomycin is the observation (VINCENT et al., 1967) that mitomycin causes a decrease in total globulin without any significant changes in the serum or in total protein.

B. Streptonigrin

1. Origin, Biological and Chemical Properties

Streptonigrin, a metabolite of *Streptomyces flocculus*, was discoverd by RAO and CULLEN (1959/196o). This antibiotic inhibits the growth of gram-positive and gram-negative bacteria (for summary see BHUYAN, 1967). Besides its cytotoxic effect, streptonigrin exhibits striking

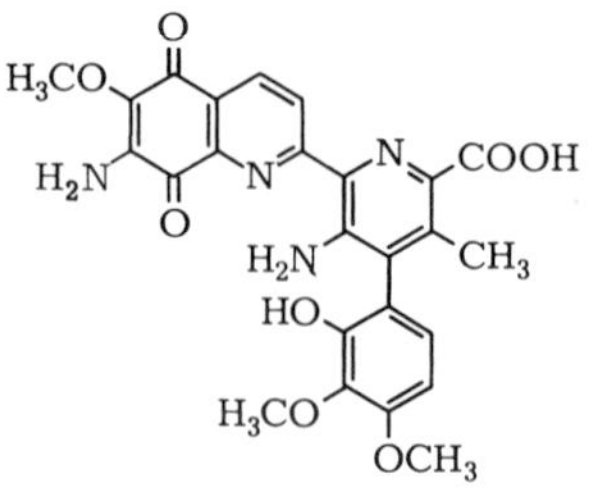

H$_3$CO

H$_2$N

N   COOH

H$_2$N   CH$_3$

HO

H$_3$CO

OCH$_3$

Fig. 5. The structure of streptonigrin. (According to RAO et al., 1963. From: Inhibitors, Tools in Cell Research, Mosbach Colloquium 1969. Berlin-Heidelberg-New York: Springer 1969, p. 413)

activity against a variety of animal tumors (OLESON et al., 1961; WILSON et al., 1961).

The structure of streptonigrin was elucidated by RAO et al. (1963). Streptonigrin (Fig. 5) is a monobasic acid and is readily susceptible to reversible two-electron reduction. The aminobenzoquinone ring A is very similar to that of mitomycin. Although these compounds are otherwise very different, both antibiotics lose activity upon replacement of the primary amino group. This strongly suggests a relationship between the aminoquinone structure in both molecules and their marked anticancer activity.

2. On the Molecular Mechanism of Action

Interaction with DNA in vitro: Streptonigrin has been shown to interact with DNA. Unlike the mitomycins, streptonigrin need not be reduced and does not cause cross-links in the DNA molecule, nor does it exhibit a selectivity for a certain base (IYER and SZYBALSKI, 1964). WHITE and WHITE (1966, 1968), however, observed that streptonigrin caused DNA degradation when chemically reduced in the presence of DNA. DNA was incubated with the antibiotic *in vitro*. No change in viscosity of the solution occurred unless streptonigrin was reduced. The authors suggested that either the conformation of the DNA molecule changes or that the molecular weigth decreases upon interaction with streptonigrin in the presence of a reducing agent. Sedimentation experiments in sucrose gradients with denatured DNA pretreated with streptonigrin and a reducing agent indicate that individual strands of DNA have been broken. The treated samples move at a rate appreciably less than that of corresponding controls. It is not clear whether a fully reduced or a semi-reduced (free radical) form of the antibiotic is the active species.

MIZUNO and GILBOE (1970) found that there must be at least two types of binding of streptonigrin to calf-thymus DNA *in vitro*: one which is reversible by dialysis and the other which is irreversible. The extent of stably bound streptonigrin was found to be 1 mole per 2,000 moles of deoxynucleotides. More streptonigrin was bound to denatured DNA or poly dC:dG, than to native DNA. It was associated preferentially with the dCMP moiety of DNA (Table 1). Reduction of streptonigrin was not required for binding. Alkaline sucrose gradient centrifugation

Table 1. According to MIZUNO and GILBOE (197o) from: Biochim. Biophys.
Acta **224**, 323-325 (197o)

a) Effect of pH on the binding of streptonigrin to DNA
(^{3}H)Streptonigrin (4o μg) and DNA (2oo μg) were incubated in 1.o ml
of 1.5 mM NaCl-o.15 mM sodium citrate which was adjusted to the de-
sired pH with HCl or NaOH. After 24 h at 4^0, DNA was isolated and
the specific activities determined

pH	Specific activity (counts/min per A_{260nm} unit)
5.o	464o ± 113
6.o	573o ± 241
7.o	736o ± 14o
8.o	736o ± 3o2
9.o	657o ± 9o

b) Specific activity of hydrolysis products of streptonigrin-DNA

5'-Deoxynucleotide	Specific activity (counts/min per μmole)
dCMP	138o
dGMP	93o
TMP	528
dAMP	39o

showed evidence of single-strand breaks in streptonigrin-treated DNA.
An increased denaturated state was indicated by hyperchromicity at
26o nm and increased buoyant density by isopycnic centrifugation in
CsCl.

Streptonigrin was tested for its effect on RNA polymerase activity
(*E. coli* enzyme) and on DNA polymerase I activity by MIZUNO (1965).
The antibiotic inhibits *in vitro* both the synthesis of RNA and the synthesis of DNA to the same extent. The inhibitory effect could be reversed by the addition of DNA. It was therefore suggested that streptonigrin and/or its metabolites may bind to the DNA helix in such a
manner that its template function for both RNA and DNA polymerase activities is impaired. The DNA polymerase I repairs DNA. Therefore
no conclusions can be drawn on the effect of streptonigrin on DNA
replication *in vivo* from the *in vitro* experiments.

Effect on DNA synthesis in whole cells: MIZUNO (1965) studied the binding
of streptonigrin to different cell components after incubation of
ascites tumor cells with ^{3}H-streptonigrin. Isolated DNA from these
cells contained about 1o times more radioactivity than RNA or protein.
Experiments with synchronized tissue-culture cells revealed that
streptonigrin was bound preferentially to DNA during the S (DNA synthetic) period of the cell cycle (MIZUNO and GILBOE, 197o). Streptonigrin inhibited totally the synthesis of DNA in *Salmonella typhimurium*
cells (LEVINE and BOTHWICK, 1963a) at a concentration of 2 x $1o^{-5}$M
and an incubation time of 9o min. Under these conditions the synthesis of RNA was inhibited by 47% and protein synthesis by 57%. In

ascites tumor cells, MIZUNO found that streptonigrin inhibited both
DNA and RNA synthesis to the same extent, indicating that the preferential effect of the antibiotic for either process may depend on
the organism. WHITE and WHITE (1968) showed that streptonigrin is
lethal to cultures of E. coli at concentrations that allow DNA, RNA
and protein synthesis to proceed. At higher concentrations DNA synthesis is preferentially inhibited. The lethal effect is accompanied
by DNA degradation, an effect which is enhanced if protein synthesis
is inhibited by withholding a required amino acid or by adding chloramphenicol. In this respect streptonigrin behaves exactly like mitomycin. The degradation products include nucleotides and bases, but
no unusual products are detected. Both an electron source and oxygen
are required for streptonigrin to exert its greatest lethal effect,
which suggests that a reaction product of oxygen together with intracellular reduced streptonigrin is lethal. Hydrogen peroxide is excluded as a likely candidate.

A quite different mode of action is discussed by HOCHSTEIN et al.
(1965), who showed that streptonigrin catalyzes the oxidation of intra-
mitochondrial NADH or NADPH with diaphorase from the soluble fraction
of the cell. Unlike the vitamin K_3-dependent oxidation of NADH or
NADPH mediated by the same enzyme, streptonigrin-induced oxidation
is not coupled to phosphorylation, but leads instead to the generation
of hydrogen peroxide. It is suggested that the toxicity of streptonigrin results from one or more of the following effects: 1. depletion of cellular NADH or NADPH, 2. uncoupling of phosphorylation followed by depletion of cellular ATP, and 3. formation of peroxide. It
is suggested that the effects of streptonigrin on cellular biosynthetic processes are consequences of its profound action on electron
transport and the associated formation of hydrogen peroxide.

3. Particular Activities in Biological Systems

Like mitomycin, streptonigrin induces phage production in inducible
strains, e.g. lysogenic Salmonella typhimurium. RADDING (1963) studied
the effect on E. coli K-12 (λ) and found that streptonigrin causes a
rapid degradation of bacterial DNA. The mechanism by which the induction occurs is still unknown. It seems probable, however, that mitomycin and streptonigrin exert their effect on phage production by
their interaction with DNA or their effect on DNA metabolism. The
antibiotic also enhances genetic recombination in phages (LEVINE
and BOTHWICK, 1963 b). Streptonigrin causes chromosome breakage in
human leukocytes in vitro (COHEN et al., 1963), affecting certain
regions on the chromosome. That the effect on the chromosome is related to a direct interaction with DNA or DNA metabolism is denied
by HOCHSTEIN et al. (1965), who suggest that chromosomal breakdown
may be caused by hydrogen peroxide generated when reduced streptonigrin is reoxidized intracellularly by oxygen. However, this mechanism seems unlikely in bacterial systems since, for example,
phenazone methosulfate causes the formation of much more hydrogen
peroxide in the presence of E. coli cultures than does streptonigrin,
but without appreciable lethality for the cells.

C. Sibiromycin

Sibiromycin is produced by Actinomycetes streptosporangium sibiricum (GAUSE
and DUDNIK, 1969, GAUSE et al., 197o). The antibiotic was found to
inhibit preferentially the growth of cells from ascites tumors, from

the reticulo-endothelial sarcoma of mice and from mouse lymphadenoma. High concentrations of sibiromycin also inhibit the growth of various bacteria, particularly *Bacillus mycoides*.

Sibiromycin has been obtained in pure form. The empirical formula is $C_{24}H_{34}N_3O_6$; the antibiotic contains one amino group, three C-methyl groups and two groups that can be acetylated. The compound possesses amphoteric properties and is soluble in dilute acids and alkalis (BRAZHNIKOVA et al., 1970).

In 1n hydrochloric acid sibiromycin is transformed into "the product of acidic inactivation" (PAI) $C_{24}H_{29}N_3O_6$. Hydrolysis of PAI with 6n HCl affords the product of acidic hydrolysis (PAH), $C_{16}H_{14}N_2O_3$. Methanolysis of sibiromycin and PAI yields the methylglycoside of a new amino sugar sibirosamine, $C_8H_{16}N\ O_3\ (OCH_3)$. Aqueous alkaline hydrolysis of PAI and PAH affords the crystalline substance $C_8H_9N\ O$ (BRAZHNIKOVA et al., 1972).

Sibiromycin forms a complex with DNA but not with RNA. DNA reverses the antibacterial action of the drug. When native DNA isolated from *E. coli* is added to sibiromycin in solution, the absorption maximum of the antibiotic is shifted from 31o nm to 32o nm. Neither RNA nor albumin, even at high concentrations, changes the spectrum of sibiromycin. Furthermore no alterations in the spectrum of the antibiotic were observed upon the addition of bases of nucleic acids (adenine, thymine, uracil, guanine), ribonucleosides (adenosine, guanosine, cytidine, uridine), ribonucleotides (adenylic, uridylic, cytidylic, guanylic acids) and deoxyribonucleotides (deoxyadenylic, deoxycytidylic, thymidylic, deoxyguanylic acids). The complexing of sibiromycin with DNA produces an increase in the melting temperature of DNA. DNA with high GC content binds more antibiotic than does DNA with low GC content. Mg^{2+} is involved in complex formation (GAUSE et al., 197o). Sibiromycin complexes not only with native DNA, but also with DNA denatured by heat. In the latter case, however, the complexing is much less frequent. In a native DNA from *E. coli* abouth 8.5 nucleotides bind one molecule of sibiromycin, whereas in the denatured DNA from the same source, 24 nucleotides are required to bind one molecule of sibiromycin.

Sibiromycin degrades the complexes actinomycin-DNA and olivomycin-DNA. DNA saturated with sibiromycin does not bind actinomycin (DUDNIK et al., 1971a).

Sibiromycin was found to inhibit preferentially the synthesis of DNA in *Staphylococcus aureus* 2o9. At a sibiromycin concentration of o.25 µg/ml the synthesis of DNA was inhibited by 83%, the synthesis of RNA by 57% and the synthesis of protein by only 2o%. A preferential inhibitory effect on DNA synthesis was also found in *B. subtilis* strain 168 T⁻, met⁻. DUDNIK et al. (1971a) observed that in cultures with an impaired DNA repair mechanism, the antibacterial effect of sibiromycin is increased. A sulphur-containing derivative of sibiromycin interacts with DNA and the sulphur is eliminated (DUDNIK et al., 1971b). This strongly suggests that DNA is the main target of sibiromycin action within the cell. Sibiromycin inhibits the *E. coli* RNA-polymerase reaction *in vitro* by forming a stable complex with DNA. Increasing the concentration of Mg^{2+} reverses the inhibition. DNA and RNA synthesis are inhibited equally well in Ascites tumor cells. In isolated nuclei of rat liver and in mitochondria DNA synthesis is supressed by sibiromycin (GAUSE et al., 1972).

D. Phleomycin

1. Origin, Biological and Chemical Properties

Phleomycin was isolated from *Streptomyces verticillus* by MAEDA et al. (1956). This antibiotic specifically inhibits the growth of myco-bacteria. *Bacillus subtilis* and some gram-negative bacteria are also affected (TAKITA, 1959; ISHIZUKA et al., 1966). Furthermore phleo-mycin has been shown to possess antitumor activity (BRADNER and PIN-DELL, 1962, 1965; TANAKA et al., 1963; ISHIZUKA et al., 1966).

Phleomycin was suggested to be a complex of copper-containing pro-teins (IKEKAWA et al., 1964). Several components, A, B, C, D1, D2, E, F, G, H, I, J and K were separated from this complex by column chromatography. The single components vary in biological activity. The copper-free phleomycin has almost the same antibacterial activi-ty as the copper-containing antibiotic except against mycobacteria, in which the copper-free complex was less active (TAKITA, 1959). Dehydrophleomycin D_1 was found to be identical with bleomycin B_2 (for details see bleomycins). The structures of the bleomycins were partially elucidated by TAKITA et al. (1972a, b). The complete chem-ical structures of bleomycins were presented first by UMEZAWA (1972) at the Euchem Conference on Antibiotics. Closely related to phleo-mycins and bleomycins are zorbamycin (ARGOUDELIS et al., 1971) and Y-A 56 (ITO et al., 1971).

2. On the Molecular Mechanism of Action

a) Interaction of Phleomycin with Nucleic Acids *in Vitro*

Binding of phleomycin to DNA was first demonstrated by FALASCHI and KORNBERG (1964). Phleomycin influences the thermal melting of DNA and of dAT. The temperature at the midpoint of transition (T_m) in the thermal melting of DNA was affected. The breadth of the transi-tion σ_t, defined as the difference between the temperatures at which 3o.8% and 69.2% of the transitions occur, was also affected. In the presence of phosphate buffer and Mg^{2+}, at concentrations routinely used for the DNA polymerase I reaction, the effect of phleomycin on the melting of dAT was to increase the σ_t from 1.5°C to 8.5°C. FA-LASCHI and KORNBERG suggested, that phleomycin binds to many points along the dAT chain and interferes with the cooperative melting of long lengths of the polymer and therefore with the sharp transition seen in the absence of phleomycin. In phosphate buffer, but without Mg^{2+}, the effect of phleomycin on the melting of dAT was to displace the T_m to higher temperatures with a less pronounced increase in the σ_t (Fig. 6).

It was predicted that phleomycin attaches to thymidine in DNA (PIETSCH, 1966, 1967). PIETSCH and GARRETT (1968) concluded from IR spectra of the phleomycin-DNA complex that a carbonyl group in the DNA molecule was involved in the binding. The infrared data are not sufficient to show directly the binding site of phleomycin to DNA. Another test of thymidine as a principal site of reaction of phleomycin utilizes the ability of DNA to interact with $HgCl_2$. Theoretical considerations by KATZ (1963) and experimental evidence presented by YAMANE and DAVIDSON (196o) show that Hg^{2+} reacts preferentially with TT pairs rather than with AA pairs. If mercury and phleomycin are indeed attracted by the same site of the thymidine molecule, then pretreatment of DNA with

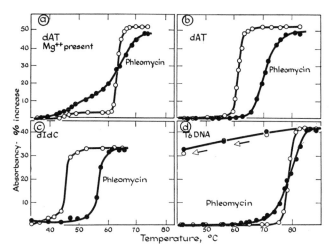

Fig. 6. Influence of phleomycin on thermal melting of DNA. (According to FALASCHI, A., and KORNBERG, A., 1964. From: Fed. Proc. 23, 943 (1964))

the antibiotic should prevent the reaction with $HgCl_2$. An interference of the interaction of Hg^{2+} and phleomycin with DNA could indeed be demonstrated. Pretreatment of DNA with either $HgCl_2$ or phleomycin prevents the other from reacting.

Although in 1969 the structure of phleomycin had not yet been elucidated, GORMAN and PIETSCH (1969) proposed a strategy for a crystallographic analysis of the phleomycin-DNA complex. Phleomycin appears to occupy discrete regions of DNA rather than to be distributed diffusely. The binding sites appear to involve a small minority of the total length of DNA (PIETSCH and GARRETT, 1969 b). Isomorphous replacement of Cu^{2+} in phleomycin by Hg^{2+} might produce diffraction data for the target sites in DNA at resolutions not yet realized with DNA fibers.

Deoxyribonucleoprotein, which binds much less actinomycin D than does pure DNA, binds phleomycin to the same extent as histone-free purified DNA, indicating that the target sites in DNA for these two antibiotics are different. From these results, and on the basis of model building, PIETSCH (1969) suggests that phleomycin, in contrast to actinomycin, is attached to DNA within the minor groove.

Direct interaction of phleomycin with reovirus RNA has been demonstrated. Phleomycin increased the temperature at the midpoint of transition in the thermal melting of this RNA by 7.2°C. Nothing is yet known about the binding sites in RNA or the functional groups in phleomycin which are involved in this type of binding (WATANABE and AUGUST, 1968).

b) Effect on DNA- and RNA-Dependent Processes *in Vitro*

Phleomycin inhibits the DNA-dependent synthesis of DNA with DNA polymerase I *in vitro* (FALASCHI and KORNBERG, 1964). The antibiotic inhibits the DNA-dependent RNA polymerase from *E. coli* to a much smaller degree. Furthermore the degradation of DNA by exonuclease I is affected by phleomycin, whereas the DNA degradation by endonuclease I is not. The inhibitory effect of phleomycin on DNA-dependent synthesis of DNA was found to be dependent on the concentration of the primer and on its content of A-T, and independent of the concentration of the enzyme.

Phleomycin, originally described as inhibiting specifically the synthesis of DNA, also interferes with RNA synthesis *in vitro* (WATANABE and AUGUST, 1968). The Qβ RNA polymerase, an RNA-dependent RNA polymerase, isolated from infected *E. coli* was used to study the *in vitro* synthesis of RNA in the presence of phleomycin. It was found that phleomycin inhibits the *in vitro* synthesis of RNA with the Qβ RNA polymerase (Fig. 7).

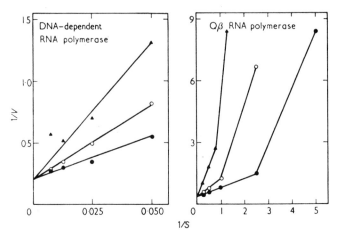

Fig. 7. Influence of template concentration on phleomycin inhibition of the DNA-dependent RNA polymerase and Qβ RNA polymerase reaction. The results are plotted according to the method of Lineweaver and Burk. Phleomycin concentrations were o.1 µg/ml (-o-o-) and o.5 µg/ml (-▲-▲-) in the DNA-dependent RNA polymerase reaction and o.2 µg/ml (-o-o-) and o.3 µg/ml (-▲-▲-) in the Qβ RNA polymerase reaction. Control samples (—●—●—) contained no phleomycin. The value of $1/V$ represents the reciprocal of GMP incorporation measured as m moles/2o min and the value of $1/S$ represents the reciprocal of template concentration (M x 10^{-6}). (According to WATANABE and AUGUST, 1968. From: J. Mol. Biol. <u>33</u>, 31 (1968))

c) Effects on Nucleic Acids and Their Metabolism in Normal and
   Virus-Infected Cells

Besides the effects of phleomycin on DNA and on RNA *in vitro*, there are several biological effects which can best be interpreted by the interaction of phleomycin with DNA within the cell. PIETSCH and COR-BETT (1968) presented direct evidence that phleomycin interacts with DNA *in vivo*. Both phleomycin and $HgCl_2$ inhibit growth presumably by interaction with DNA. In cells pretreated with phleomycin before being challenged with $HgCl_2$, the level of inhibition was less than the sum of the effects of the two agents reacted separately. The degree of protection was a direct linear function of the specific activity of phleomycin. The authors concluded that phleomycin and Hg^{2+} attack the same sites in the living cells. PIETSCH and GARRETT (1969a) isolated DNA from phleomycin-treated *E. coli* cells. Under

the electron microscope, the DNA showed nodules which were interpreted as being antibiotic molecules attached to DNA. The nodules
appeared to increase the fragility of the DNA molecule.

Like mitomycins and streptonigrin, phleomycin in cultures of bacteria
and mammalian cells inhibited DNA synthesis preferentially (TANAKA
et al., 1963; PIETSCH and McCOLLISTER, 1965; PIETSCH, 1967; KIHLMAN
et al., 1967).

Besides its effects on nucleic acid synthesis, phleomycin induces DNA
breakdown in E. coli (GRIGG, 1969). This effect is considerably ampli-
fied in the presence of caffeine (GRIGG, 197o). Spermine has some
protective effect, and it depressed phleomycin-induced DNA breakdown
by as much as 33%. The inhibitory effect of phleomycin on cell growth
could be reduced by spermine by as much as 5o%. Excisionless (HCR⁻)
mutants of E. coli are insensitive to doses of phleomycin which kill
more than 99% of the wild-type organisms within one hour. The DNA
breakdown, the inhibition of DNA replication and cell death are therefore suggested to be consequences of an initial attack on DNA by an
excision endonuclease, stimulated somehow by phleomycin. The damage
to DNA caused by phleomycin probably cannot be repaired.

PITTS and SINSHEIMER (1966) studied the effect of phleomycin on the
replication of the DNA bacteriophage φ X 174, which involves three
stages of DNA synthesis. 1. the addition of complementary strand to
single-stranded parental DNA to produce a replicative form; 2. the
semi-conservative replication of this replicative form; 3. the synthesis of single-stranded DNA for progeny phage by a conservative
process. Evidence was presented to show that phleomycin acts differently during these stages of φ X 174 DNA synthesis. Phleomycin inhibits either DNA synthesis involved in the production of the parental
replicative form (step 1) or the "replicative form" replication (step
2), while the synthesis of single-stranded DNA directed by the replicative form template is not affected.

IWATA and CONSIGLI (1971) observed that phleomycin affects the production of mature polyoma virus in mouse embryo cells. In uninfected
host cells phleomycin inhibited DNA synthesis to 96% without affecting
RNA and protein synthesis to any considerable extent. In virus-infected cells DNA synthesis proceeds, in the presence of phleomycin,
at a reduced rate compared with DNA synthesis in uninfected, untreated host cells. DNA was isolated from infected untreated or phleomycin-treated cells and fractionated by using ethidium bromide-CsCl gradients. From phleomycin-treated infected cells, twice as much circular nicked and linear DNA as supercoiled DNA was isolated than from
untreated infected cells. Viral particles produced in the presence
of phleomycin did not contain supercoiled DNA. Immunofluorescent stai-
ning studies utilizing an antibody against the viral protein showed
that all viral proteins were made in the presence of phleomycin. It
is therefore suggested that the DNA of the viral particles is defective. Since supercoils were missing it is possible that the enzymes
necessary for DNA replication such as polymerase, nickase or ligase
are not operating precisely in the presence of phleomycin.

In addition to the phleomycin effect on viral DNA formation, WATANABE
and AUGUST (1966, 1968) reported on the inhibitory action of phleomycin on Qβ phage-specific RNA synthesis. The cessation of RNA phage
replication appeared to result from an inhibition of phage-directed
RNA synthesis. After addition of phleomycin there was simultaneous
cessation of the synthesis of both single-stranded phage RNA and
RNAase-resistant RNA. The synthesis of coat protein, phage assembly
and the formation of infective particles were unaffected when phleomycin was used at concentrations which reduced only the yield of
phages. This treatment did not result in defective particles. It is

important to note, however, that maximum inhibition occurred when phleomycin was present during the first 1o min after infection. When phleomycin was added 3o min after infection, the yield of intracellular phage was almost equal to that of untreated cultures. These results strongly indicate that late processes of phage maturation were unaffected by phleomycin. When phage RNA was used as a messenger for cell-free synthesis of protein, phleomycin did not inhibit the synthesis of phage protein. HECHT and SUMMERS (197o) obtained similar results in investigations of polio virus RNA replication in HeLa cells. The normal synthesis of polio virus RNA in HeLa cells has been shown to have two phases: an early phase during which the synthesis of RNA is increasing although infectious viruses are not produced and a later phase during which the rate of RNA synthesis is constant and mature virus particles are formed. Phleomycin acted only during the early phase of RNA synthesis. It is speculated that phleomycin may inhibit polio-virus RNA synthesis by blocking the synthesis of virus specific polymerase, or that it interferes with the synthesis of other virus specific proteins. The greater effect of the drug in the early phase may be explained as follows: during this phase the pool of virus-specific RNA polymerase is much lower than later in the infection cycle. Phleomycin does not prevent the polio virus RNA polymerase complex from synthesizing RNA *in vitro*. When intact polio viruses were exposed to phleomycin at high concentrations (2oo µg/ml), the infectivity of the viruses was not impaired (KOCH, 1971). Single-stranded polio virus RNA was inactivated by this treatment by up to 99% at concentrations of 2 µg/ml whereas the infectivity of double- and multistranded RNA was 1o times less sensitive than that of single-stranded RNA. These results demonstrate the direct interaction of phleomycin with single-stranded polio virus RNA and point to the biological significance of this interaction. It is suggested that phleomycin binds to the single-stranded 3' end of the virus RNA.

d) Particular Activities in Biological Systems

A consequence of the interaction of phleomycin with DNA is the induction of chromosome- and chromatid-type aberrations (MATTINGLY, 1966; KIHLMAN et al., 1967). KAJIWARA et al. (1966) investigated the possible relationship between inhibition of DNA synthesis and the prevention of HeLa cell division. The effects of the agent, namely blockage of cells entering mitosis, were observed in cultures in which almost all the DNA had been replicated, and were obtained with concentrations that affected DNA synthesis only slightly. Kinetic studies indicated that this phleomycin-sensitive step followed completion of DNA synthesis. Microscopic evidence suggested that the agent prevented the cells from entering prophase. As was previously discussed for mitomycin, the mechanism by which phleomycin inhibits mitosis is not yet clear (DJORDJEVIC and KIM, 1967; HOTTA and STERN, 1969).

E. Bleomycin

1. Origin, Biological and Chemical Properties

Bleomycin was discovered by UMEZAWA et al. (1966 a) and further separated into each of the bleomycins A_1-A_6 and B_1-B_5 by CM-Sephadex C-25 column chromatography (UMEZAWA et al., 1966 b). The bleomycins were

25

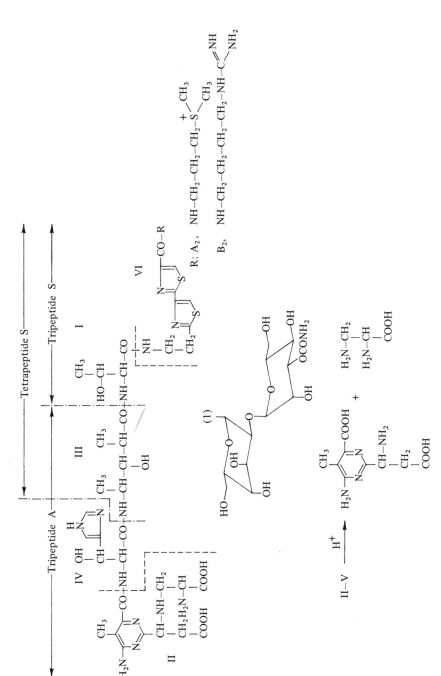

Fig. 8. Partial structures of bleomycin A$_2$ and B$_2$. (According to UMEZAWA, 1972. Manuscript of the first presentation at the Euchem Conference on Antibiotics, August 1972 at Aarhus, Denmark)

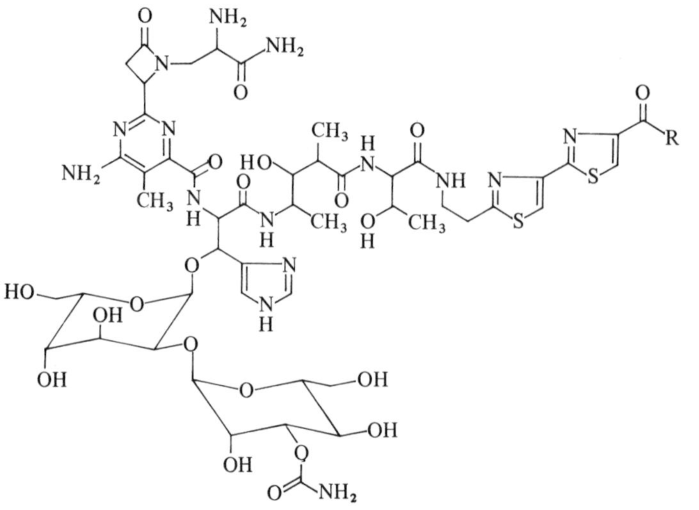

Bleomycin: R =—NH$_2$

Bleomycinic acid: R=OH

A$_2$: R=NHCH$_2$CH$_2$CH$_2$S$^+$(CH$_3$)$_2$

B$_2$: R=NHCH$_2$CH$_2$CH$_2$CH$_2$NHC(=NH)NH$_2$

Fig. 9. Structure of bleomycin. (According to UMEZAWA, see Fig. 8)

found to possess antimicrobial and carcinostatic properties (ISHIZUKA et al., 1967).

The structures of bleomycins and phleomycins have been elucidated in various laboratories (TAKITA et al., 1968, 1969, 197o, 1971, 1972a,b; KOYAMA et al., 1968; MURAOKA et al., 197o, 1972; OMOTO et al., 1972). The partial structures of bleomycin A$_2$ and B$_2$ are shown in Fig. 8. Upon mild hydrolysis of bleomycin A$_2$, two tripeptides, S and A, were obtained. Further hydrolysis of the tripeptide A resulted in four amino acids: β-amino-β-(4-amino-6-carboxy-5-methylpyrimidin-2-yl) propionic acid (II), L-β-aminoalanine (V), β-hydroxyhistidine (IV) and 4-amino-3-hydroxy-2-methyl-n-valeric acid (III). The structures of the amino acids were confirmed by total synthesis. The hydrolysis of the tripeptide S results in L-threonine (I), 2'-(2-aminoethyl-)-2,4'-bithiazole-4-carboxylic acid (VI) and 3-aminopropyl-dimethylsulfonium (VII). The amino acid (VI) was synthesized by ZEE-CHENG and CHENG (197o). With N-bromosuccinimide, mild hydrolysis resulted in tetrapeptide S. Besides the unusual amino acids, the antibiotic contains a disaccharide consisting of a methylglycoside of L-gulose and 3-O-carbamoyl-D-mannose. All bleomycins and phleomycins contain the same sugar moiety. On the basis of these findings the complete structures of bleomycin A$_2$ and B$_2$ were proposed (Fig. 9). The bleomycins are thus peptides containing a β-lactam ring, a pyrimidine moiety and a sugar moiety. The acid moiety without the amine is called bleomycinic acid and is common to all bleomycins.

Total hydrolysis of phleomycin D$_1$ gave the same amino acids (and agmatine) as were obtained on hydrolysis of bleomycin B$_1$ except for the bithiazole acid, which is replaced by β-aminoalanine and 2-acetyl-

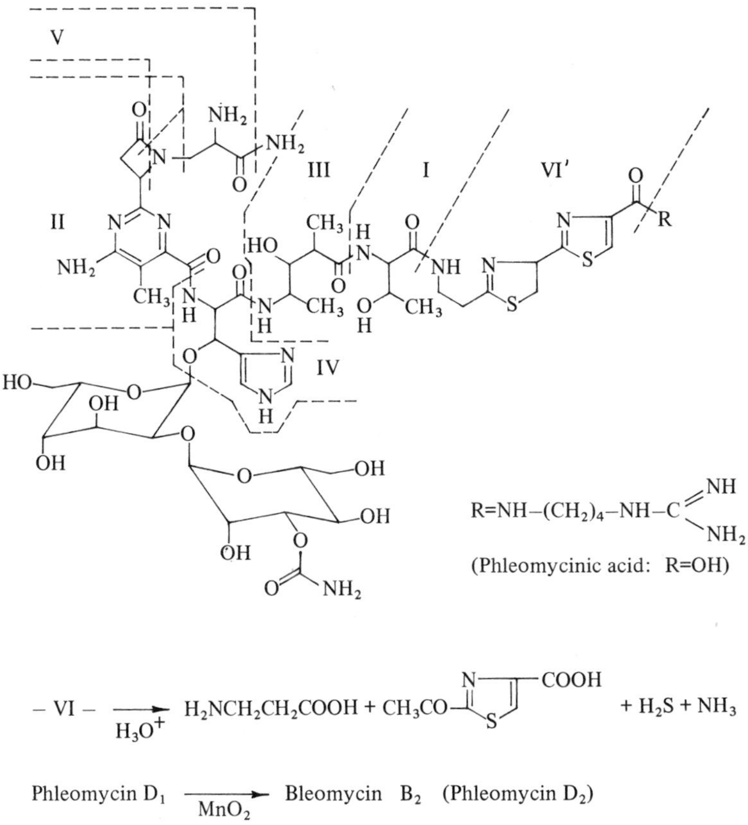

Fig. 1o. Structure of phleomycin D_1. (According to UMEZAWA, see Fig. 8)

thiazole-4-carboxylic acid. By oxidation with MnO_2, phleomycin D_1 was transformed to bleomycin. The proposed structure for phleomycin D_1 is shown in Figur 1o. In analogy to bleomycinic acid, the amine-free compound is called phleomycinic acid.

2. On the Molecular Mechanism of Action

a) Interaction with DNA *in Vitro*

Bleomycin causes strand scission of both native and double-stranded DNA of *E. coli* or HeLa cells, but has no effect on ribosomal and transfer RNAs under the same conditions. The strand scission caused by the antibiotic is inhibited by the addition of Cu^{2+}, Co^{2+}, Zn^{2+} and EDTA (SUZUKI et al., 197o). In previous studies the effect of bleomycin on the scission of DNA strands was only observed in the presence of hydrogen peroxide (NAGAI et al., 1969 a) or in the presence of 2-mercaptoethanol (NAGAI et al., 1969 b). More detailed studies by SHIRAKAWA et al. (1971) indicated that strand scission of DNA by bleomycin also occurs in the absence of sulfhydryl- or peroxide compounds.

Furthermore SUZUKI et al. (197o) demonstrated direct binding of radioactively-labeled bleomycin to DNA. When Sephadex column chromatography was used to isolate the complex, the molar ratio of DNA nucleotide to bleomycin was found to be 35o:1, whereas equilibrium dialysis
gave a molar ratio of 15:1. This difference in the number of bleomycin
molecules bound to DNA is explained as being caused by a release of
bound bleomycin from DNA during the elution process of Sephadex column chromatography. By using this chromatographic method it was shown
that double-stranded DNA binds less bleomycin than heat-denatured
DNA.

The results can be interpreted by assuming that the binding forces
to native DNA are much weaker than those for heat-denatured DNA. The
divalent cations Zn^{2+}, Co^{2+} and Cu^{2+}, - which hinder DNA strand scission by phleomycin - also prevent the binding of the bleomycin to
DNA (SUZUKI et al., 197o). It is suggested that bleomycin reacts directly with these metal ions, because it has chelating properties.
Bleomycin A_2 causes strand scission also in poly (dG), poly (dC) and
poly d(AT) in the presence of 2-mercaptoethanol. HAIDLE (1971) reported the fragmentation of DNA from *Bacillus subtilis* by bleomycin in
the presence of 2-mercaptoethanol.

In contrast to phleomycin, bleomycin decreases the T_m of DNA (*E. coli*,
salmon sperm, NAGAI et al., 1969 a, b). The differences in the interaction of bleomycin and phleomycin with DNA were attributed to the
presence of a different amino acid in phleomycin D_1, see Figure 8 VI
(2-[2-(2-aminoethyl)-Δ^2-thiazolin-4-yl]-thiazole-4-carboxylic acid).

That direct interaction of bleomycin with DNA results in a function-
ally- and structurally-altered DNA is shown in the experiments of
SHIRAKAWA et al. (1971): The infectious properties of isolated single-
stranded circular DNA of phage ϕ X 174 for *E. coli* spheroblasts is lost
after exposure to bleomycin. The interaction of bleomycin with phage
DNA leads to a decrease in sedimentation velocity of the phage DNA.
The bleomycin effects are enhanced by sulfhydryl compounds and hydrogen peroxide and counteracted by Cu^{2+}.

Interaction of DNA with zorbamycin, phleomycin and bleomycin was
studied using UV absorption and circular dichroism measurements by
KRUEGER et al. (1973). The physical measurements indicate, that these
antibiotics definitely bind to DNA.

b) Effect on DNA-Dependent Processes *in Vitro*

MIYAKI et al. (1971) observed that bleomycin A_2 inhibited the ATP
ligase reaction by an enzyme preparation obtained from rat ascites
hepatoma 13o cells. The action of a DNA ligase prepared from T_4 phage-
infected *E. coli* B on *E. coli* DNA nicked by pancreatic DNAase was also
inhibited by bleomycin A_2 (TERASIMA et al., 197o). These findings are
in good agreement with the postulation of IWATA and CONSIGLI (1971)
that the enzymes necessary for DNA replication such as polymerase,
nickase or ligase do not function properly in the presence of phleomycin. MÜLLER et al. (1972 a) reported that bleomycin inhibited DNA-
dependent DNA polymerase obtained from Rauscher murine leukemia virus
in the presence of dithiothreitol. Concentrations of these antibiotics a hundred times higher were required to inhibit a RNA-dependent
DNA polymerase from the same source.

The action of bleomycin on enzymatic DNA, RNA and protein synthesis
in assays using DNA-dependent DNA polymerase and DNA-dependent RNA

polymerase from *E. coli* was later investigated by MÜLLER et al. (1972 b) and YAMAZAKI et al. (1973) with DNA rich in d Ado and d Thd and by d Guo, d Cyd rich DNA. Bleomycin inhibits the synthesis of nucleic acids with d Ado and d Thd rich DNA but conversely stimulates the synthesis of nucleic acids with the d Guo d Cyd rich primer. Evidence is presented, that the initiation step is the primary target for bleomycin stimulation as well as for inhibition. Bleomycin A_2 stimulated pancreatic DNAase activity (SHIRAKAWA et al., 1971).

Besides this stimulatory effect, which may occur with a certain template the opposite observation was made by MÜLLER et al. (1973). Bleomycin inhibits strongly DNAase I and much less DNAase II. This is in good agreement with the characteristics of DNAase II which favours the cleaveage of dGp-Cp linkages in DNA. RNAase A, B, ribonuclease T_1, phosphodiesterase I and II are not affected by bleomycin.

c) Effects on Nucleic Acids and Their Metabolism within the Cell

Bleomycin inhibits the incorporation of ^{3}H-thymidine into DNA - an effect which was observed in cultured mammalian cells. In addition to this, bleomycin causes degradation of DNA (TERASIMA et al., 197o).

TERASIMA and UMEZAWA (197o) observed that the effect on HeLa cells is dependent on the phase of the growth cycle. The cells are most sensitive to bleomycin during mitosis (BARRANCO and HUMPHRY, 1971). Therefore, the primary biological action of bleomycin is assumed to interfere with transcriptional or translational events involved with the synthesis of division-specific protein(s) which is presumably produced at the G_2 phase. This assumption agrees well with the observations of FUJIWARA et al. (manuscript in preparation) who showed in experiments with human lymphocyte cultures that both short (4 h) and prolonged (12 h) exposure to bleomycin, before chromosome preparation, induce chromosome aberrations nonrandomly and to a similar extent at a 1o% level of the cells which can enter mitosis successfully. In HeLa S 3 cells bleomycin has only little effect. Only high concentrations exert some inhibitory effect on the joining of short segments of replicating DNA after a 3o-min ^{3}H-thymidine pulse, but the joining ability is soon resumed. The data suggest, that bleomycin may either hardly enter HeLa S 3 cells or may be readily inactivated (FUJIWARA and KONDO, 1973).

ENDO (197o) reported that *E. coli* mutants sensitive to UV only or UV and X-rays were equally as sensitive to bleomycin as was the isogenic resistant strain. The data of ENDO support the view that the effect of bleomycin is different from that of the so-called radiomimetic agents and of X-rays. If bleomycin affects the ligase it is probable that the excision-repair mechanism will not function in the presence of bleomycin.

A recombination deficient mutant of *B. subtilis* (rec⁻ strain) behaved differently in the presence of bleomycin than did the respective wild type organism. The wild type cells were able to recover from lower levels of bleomycin while the rec⁻ strain was not. A bleomycin resistant strain of *B. subtilis* was isolated and found to be cross-resistant to mitomycin but not to daunomycin (SAUNDERS and SCHULTZ, 1972).

Bleomycin causes the rapid induction of both λ bacteriophage in *E. coli* K_{12} (λ) and the defective bacteriophage PBSH in *Bacillus subtilis* 168 and insofar shares some biological properties with mitomycin and alkylating agents (HAIDLE et al., 1972).

F. Neocarcinostatin

1. Origin, Biological and Chemical Properties

Neocarcinostatin is produced by *Streptomyces carcinostaticus*. This antibiotic shows antibacterial activity only against a few gram-positive
bacteria; on the other hand, it exhibits marked antitumor activity
against various mouse and rat ascites tumors (ISHIDA et al., 1965).
The biological activity of neocarcinostatin is altered to a significant degree by deamination, particularly in its action towards
bacteria (KUMAGAI et al., 1966).

Neocarcinostatin is an acidic protein with a molecular weight of
about 9,ooo (MAEDA et al., 1966). The deaminated compound was prepared from neocarcinostatin with $NaNO_2$ under acidic conditions. It
possesses the lysine-ε-aminogroup of the original neocarcinostatin,
but lacks the N-terminal NH_2-group of alanine. The chemical properties of neocarcinostatin were studied by optical rotation measurements by MAEDA and ISHIDA (1967). These investigations revealed that
neocarcinostatin was free of α-helix-type structures. The amino acid
residues seryl-, threonyl-, glycyl-, valyl- and prolyl-present in
this protein make up approximately 5o% of the total amino acids. It
is suggested that neocarcinostatin possesses fixed tyrosyl and phenylalanyl residues, which contribute significantly to the optical rotation above 26o mμ. The deaminated neocarcinostatin seems to be altered to some extent in its conformation, allowing free rotation of
the phenylalanyl side chain residue. The difference between the biological activities of the neocarcinostatin and its deaminated form
may thus be attributed to a conformational change as well as to a
chemical change. The complete chemical structure (Fig. 11) was elucidated by MEIENHOFER et al. (1972). The protein contains 1o5 amino
acids with alanine residues at the amino terminal and asparagine at
the carboxyl terminal.

Ala–Ala–Pro–Thr–Ala–Thr–Val–Thr–Pro–Ser–Ser–Gly–Leu – Ser–
           5   10
Asp–Gly–Thr–Val–Val–Lys–Val–Ala–Gly–Ala–Gly–Leu–Gln – Ala–
15   20   25
Gly–Thr–Ala–Tyr–Asp–Val–Gly–Gln–Cys–Ala–Ser–Val–Asn –Thr–
       30   35   40
Gly–Val–Leu–Trp–Asn–Ser–Val–Thr–Ala–Ala–Gly–Ser–Ala –Cys–
       45   50   55
Asx–Pro–Ala–Asn–Phe–Ser–Leu–Thr–Val–Arg–Arg–Ser–Phe –Glu–
           60   65   70
Gly–Phe–Leu–Phe–Asp–Gly–Thr–Arg–Trp–Gly–Thr–Val–Asx–Cys–
           75   80
Thr–Thr–Ala–Ala–Cys– Gln–Val–Gly–Leu–Ser–Asp–Ala–Ala–Gly–
85   90   95
Asp–Gly–Glu–Pro–Gly–Val –Ala–Ile– Ser– Phe–Asn
       100   105

Fig. 11. Primary structure of neocarcinostatin. (According to MEIEN-
HOFER et al., 1972)

2. On the Mechanism of Action

Neocarcinostatin is a very effective inhibitor of DNA synthesis in bacteria, e.g. *Sarcina lutea* (ONO et al., 1966) as well as in HeLa cells in culture (HOMMA et al., 197o). In *S. lutea* DNA synthesis was inhibited immediately after the addition of neocarcinostatin at concentrations as low as o.oo5 µg/ml. The synthesis of RNA and of protein was not inhibited up to concentrations of o.5 µg/ml, but the DNA was markedly degraded into free bases. Although considerable DNA degradation occurred, the DNA extracted from treated cells was able to serve as a template for DNA synthesis *in vitro* and did not differ from normal DNA in the thermal denaturation profile. The effect of neocarcinostatin on DNA synthesis and breakdown resembles to some extent that of mitomycin. This view is supported by the observation that neocarcinostatin, like mitomycin, is able to induce the development of active phages in lysogenic strains of *E. coli* (PRICE et al., 1964; HEINEMANN and HOWARD, 1964).

HOMMA et al. (197o) studied the effect of neocarcinostatin in non-synchronized cultures of HeLa cells and found that mitosis was affected immediately after the addition of the antibiotic. Although DNA synthesis was also immediately inhibited, eight hours were required to produce the maximal effect. This period corresponds exactly to the period of DNA synthesis in the HeLa S_3 cells (S phase). Kinetic measurements of the inhibition of DNA synthesis were also made in synchronized cultures: The addition of neocarcinostatin at any time during the S phase did not inhibit the synthesis of DNA. Neocarcinostatin does not affect the cells which are passing through the S phase, but it affects the cells which are in the G_1 phase or just entering the S phase.

If one compares the effect of neocarcinostatin with that of phleomycin on HeLa cells, it is obvious that both antibiotics inhibit DNA synthesis and block mitosis. In contrast to neocarcinostatin, phleomycin inhibits cell mitosis at doses which do not affect DNA synthesis (KAJIWARA et al., 1966). Phleomycin inhibits the activity of DNA polymerase I by direct interaction with the primer DNA (FALASCHI and KORNBERG, 1964), whereas neocarcinostatin does not affect DNA polymerase I activity *in vitro*.

These results indicate that the inhibition of DNA synthesis, caused by this antibiotic, does not involve a direct interaction with DNA or with the Kornberg enzyme. The possibility remains that the polypeptide interferes with membrane-bound DNA-synthesizing enzyme systems. Neocarcinostatin is highly specific with regard to its inhibitory effect on DNA synthesis and the concomitant blockage of mitosis. Thus one might speculate that this inhibitor interferes with the enzymes responsible for DNA replication *in vivo*.

G. Edeine

Edeines are structurally related oligopeptide antibiotics produced by *Bacillus brevis*. They inhibit gram-negative and gram-positive microorganisms (KURYLO-BOROWSKA and SZER, 1972). The major representative of this group is edeine A, which is referred to as edeine.

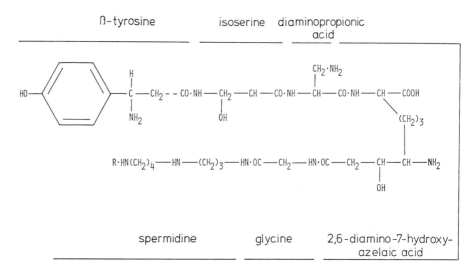

Fig. 12. Structures of edeines. Edeine A_1, R is H; edeine B_1, R is C-NH; edeine A_2 and B_2, isoserine residue linked to β-amino group of α,β-diaminopropionic acid. (According to HETTINGER, T.P., and CRAIG, L.C., 197o)

The structures of the edeines were elucidated by HETTINGER et al. (197o) and HETTINGER and CRAIG (197o; see Fig. 12). The oligopeptide antibiotic contains spermidine, glycine, 2,6-diamino-7-hydroxyazelaic acid, diamino-propionic acid, isoserine and β-tyrosine. Besides these components 3-amino-2-hydroxy-propionic acid and 3-amino-3-phenyl-propionic acid were found (WOJCIECHOWSKA et al., 1972).

The antibiotic appears to be a specific inhibitor of bacterial DNA synthesis; it does not interfere with the growth and burst size of bacteriophage T_4 (TABACZYNSKI and JABLONSKA, 197o). In vitro studies with purified E. coli DNA polymerase I showed that edeine does not interfere with the DNA polymerase I reaction in vitro. KURYLO-BOROWSKA and SZER (1972) examined the effect of edeine on isolated DNA-membrane complexes which carry out semi-conservative DNA synthesis on endogenous templates. The membrane complexes were derived from a mutant which lacks or contains very low levels of DNA polymerase I.

Edeine was found to inhibit in vitro DNA synthesis in these membrane complexes with endogenous templates. The active inhibitor is edeine A_1, indicating that the effect of edeines on in vitro systems capable of replicating DNA parallels their in vivo effect.

Edeines are basic pentapeptides containing a spermidinyl residue at the carboxyl terminal. Accordingly, edeine A_1 was tested with respect to its specificity. Comparative experiments with spermine and spermidine showed that the polyamines inhibit DNA polymerase reaction with the DNA-membrane complex at relatively high levels, whereas edeine exhibits its effect at concentrations of about $1o^{-6}$ M.

In cell-free systems of protein synthesis, the antibiotic blocks ribosomal binding sites for aminoacyl tRNA. Edeine will become a useful tool in the study of replicating systems of DNA synthesis which are free of DNA polymerase I.

H. Nalidixic Acid

1. Origin, Biological and Chemical Properties

Nalidixic acid, a synthetic compound with antibiotic properties, was synthesized by the Sterling-Winthrop Laboraties, USA, and first described by LESHER et al. (1962). Nalidixic acid and several derivatives were found to be highly effective antibacterial agents, both *in vitro* and *in vivo*. The *in vivo* activity of these compounds is most pronounced against gram-negative bacteria, while gram-positive organisms are generally more resistant (Table 2).

Table 2. Antibacterial activity of nalidixic acid

Microorganism	Minimal bacteriostatic concentration, mcg/ml	ED_{50} (mg/kg) in mice.p.o.
Escherichia coli	5.o - 12.5	<25
Pasteurella spp.	o.5 - 2.5	<25
Klebsiella pneumoniae	o.8 - 25.o	6o
Aerobacter aerogenes	1.6 - 25.o	35
Proteus spp.	1.25- 3o.o	5o
Salmonella spp.	3.2 - 5o.o	62
Shigella spp.	o.8 - 3.2	<25
Brucella spp.	7.5 - 1o.o	>4oo
Staphylococcus aureus	5o.o -1oo.o	>4oo
Diplococcus pneumoniae I	25o	>4oo

According to LESHER et al. (1962), from: J. Med. Pharm. Chem. 5, 1o64.

Nalidixic acid belongs to a class of chemotherapeutic agents which can be described as 1,8-naphthyridine derivatives. The derivatives of 1,8-naphthyridine are shown in Fig. 13. The outstanding compound of this series is 1-ethyl-7-methyl-1,8-naphthyridine-4-one-3-carboxylic acid, which is called nalidixic acid.

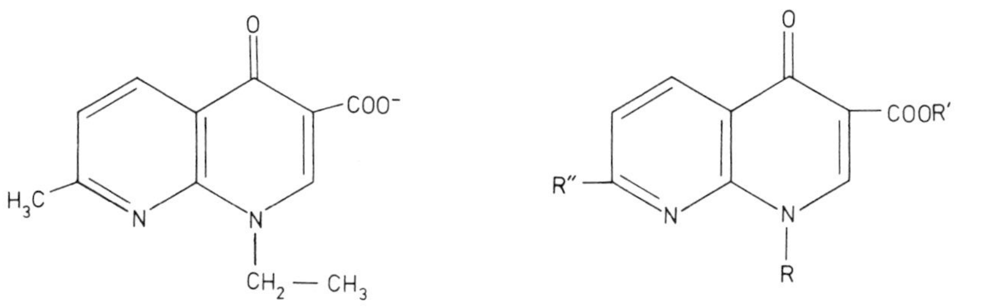

Nalidixic acid

1-ALKYL-1,8-NAPHTHYRIDIN-4-ONE-3-CARBOXYLIC ACID DERIVATIVES

Fig. 13. Structure of nalidixic acid and derivatives of 1,8-naphthyridine. (According to LESHER et al., 1962. From: J. Med. Pharm. Chem. 5, 1o63 (1962))

2. Mechanism of Action and Biological Activities

a) Effect on DNA and Nucleic Acid Metabolism

The mechanism of action of nalidixic acid was studied in gram-negative and gram-positive microorganisms and in mammalian cell cultures.
GOSS et al. (1964, 1965) observed that nalidixic acid, even at low
levels, exerted a bactericidal effect on *E. coli*. It was suggested
that the development of extremely long filamentous cells with concomitant loss of viability resembles the effect caused by thymine
deficiency. The lethal effect on mammalian cells is manifested by
an arrest of proliferation and accompanied by morphological alterations of susceptible cells.

Chemical analysis of cellular constituents provided direct evidence
that nalidixic acid acts preferentially on DNA. ROSENKRANZ and LAMBEK
(1965) isolated a DNA from treated bacteria which was reversibly de-
naturable, indicating cross-linkage between the polynucleotide strands.
They suggested that the action of nalidixic acid is closely related
to the effect of nitrogen mustard or mitomycin. Moreover the "cross-
linked DNA" isolated from nalidixic acid-treated bacteria was unstable
in solution - as would be expected if the effect were caused by alkylating agents. ROSENKRANZ and LAMBEK (1965) used DNA from a line of
stable human amnion cells for thermal denaturation and renaturation
studies. The analysis of bouyant density of DNA showed structural
modifications. However, no direct evidence has yet been obtained to
show that.

Nalidixic acid has been studied for its action on purified enzymes
acting on DNA and in subcellular DNA-synthesizing systems (PEDRINI
et al., 1972 b). No inhibition was found for DNA polymerase I, endonuclease I, exonuclease I, II and III from *E. coli*, polynucleotide-
ligase and DNA methyl-transferase from T_4 infected *E. coli*, and DNA
polymerase from *B. subtilis*. A significant inhibition however of ATP
dependent synthesis was observed under certain conditions in tolue-
nized *E. coli* strains lacking DNA polymerase I. PEDRINI and coworkers
conclude a mode of action through a still unidentified physiological
component of the growing point apparatus.

COOK et al. (1966a) reported the effects of nalidixic acid on the
stability of cellular constituents. They showed that treatment of
E. coli 15TAU⁻ with nalidixic acid resulted in degradation of DNA and
of RNA, while proteins were unaffected. DNA degradation appeared to be
more extensive than RNA degradation during periods of comparable bactericidal action. The onset of DNA degradation was evident prior to
a measurable bactericidal effect. Within the range of 2-2o% DNA degradation, however, a decrease in viable cell number was found. Degradation of DNA to acid-soluble material occurred only under conditions permitting the bactericidal action of nalidixic acid. DNA degradation, however, seemed not to be an inevitable consequence of the
action of nalidixic acid. Using the *E. coli* 15TAU⁻ mutant, it was demonstrated that the primary inhibition of DNA synthesis could be separated from the secondary bactericidal effect. Although DNA synthesis in *E. coli* 15TAU⁻ is blocked by nalidixic acid in the (+T, -AU)
medium, there is no detectable loss of viability and essentially no degradation of DNA; as in the case of mitomycin some RNA degradation
was found.

Mitomycin and nalidixic acid differ in their effects on DNA degradation. Whereas chloramphenicol enhances the mitomycin-induced DNA de-

gradation, the inhibitor of protein synthesis restricts DNA degradation and loss of viability in nalidixic acid-treated cells.

The conditions required for lethality were studied by DEITZ et al. (1966), who found that in *E. coli*, the inhibition of DNA synthesis was proportional to the concentration of nalidixic acid, even at sublethal doses. Removal of the drug from treated cultures resulted in immediate restoration of DNA synthesis; retreatment caused inhibition of DNA synthesis. This finding suggests that nalidixic acid is not irreversibly bound to sensitive cellular sites. It is, however, possible that a repair mechanism is involved in recovery. The cells appear to be capable, once the drug is removed, of rapidly repairing the damaged site.

Similar results on the effect of nalidixic acid on DNA synthesis were obtained with *B. subtilis* (COOK et al., 1966c). RAMAREDDY and REITER (1969) described specific loss of newly replicated deoxyribonucleic acid in nalidixic acid-treated *B. subtilis* 168. The finding that in proliferating cell cultures the inhibition of DNA synthesis is followed by extensive loss of DNA, led these authors to explore the mode of DNA degradation further. They found that the effect of nalidixic acid resembles that of thymine starvation. The degradation of DNA caused by nalidixic acid appears to be initiated at the replication point and to proceed sequentially along the bacterial chromosome from the most recently synthesized DNA to "older" DNA. Both DNA strands appear to be degraded equally, and essentially all of the products of degradation are soluble in cold acid. It is suggested that nalidixic acid causes the degradation of DNA just behind the replication point. Apparently any agent that selectively arrests DNA synthesis in an actively replicating cell might alter the complex of nuclease-polymerase-ligase activity. When triphosphates were omitted from a preparation of the polymerase complex+DNA, the nuclease activity of this complex was greatly enhanced (GANESAN, 1968). The author further discusses the probability that the DNA inhibitor acts primarily to release prophages, and that DNA degradation is a secondary event brought about by a specific nuclease produced as a consequence of phage induction. One can, however, argue equally well from the opposing viewpoint, that degradation of cell DNA is the primary response to these agents and that phage induction is a consequence of degradation of DNA near the site of prophage integration.

GAGE and FUJITA (1969) and PEDRINI et al. (1972 a) studied the effect of nalidixic acid on deoxyribonucleic acid synthesis in *B. subtilis* infected with bacteriophage SPO1. Nalidixic acid had little inhibitory effect on SPO1 DNA synthesis at concentrations that drastically inhibited *B. subtilis* DNA synthesis. GAGE and FUJITA report that the SPO1 DNA synthesized in the presence of high concentrations of nalidixic acid had a density characteristic of normal SPO1 DNA and was packaged into viable progeny phage particles, but its rate of synthesis was reduced and bacterial lysis was delayed (Fig. 14).

Since nalidixic acid appears to enter infected cells as rapidly as it does uninfected cells, and has the same effect on DNA synthesis whether it is added before or after infection, it seems unlikely that a differential effect of the drug can be attributed to a post-infection change in cell-wall or membrane permeability or to a difference in intracellular drug concentrations. Active catabolism of the drug during infection also seems unlikely because the reduced rates of DNA synthesis, once established, remained constant.

SPO1 DNA contains hydroxymethyl uracil instead of thymine. Therefore, in the course of normal infection, alterations are made in the *B. sub-*

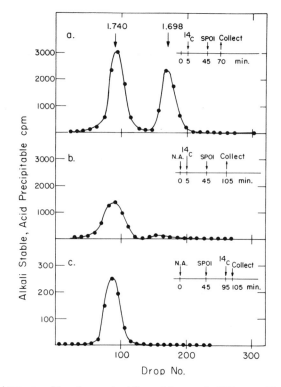

Fig. 14. Density gradient centrifugation of DNA synthesized in the presence of nalidixic acid. A culture of *B. subtilis* HA1o1 cells was divided into three parts and two (*b*, *c*) received 5o µg of nalidixic acid per ml. After 5 min, adenine-8-^{14}C (15 µg/ml, o.133 µCi/µg) was added to cultures *a* and *b*. After an additional 4o min, SPO1 (multiplicity of infection = 1o) was added to all three. (*a*) Control, no nalidixic acid, cells collected 25 min after infection; (*b*) 5o µg of nalidixic acid per ml, cells collected 6o min after infection; (*c*) 5o µg of nalidixic acid per ml, adenine-8-^{14}C (6 µg/ml, o.33 µCi/µg) was added at 5o min after infection, and the cells were collected at 6o min after infection. Refractometrically determined densities in CsCl are shown over arrows. (According to GAGE and FUJITA, 1969. From: J. Bact. **98**, 1o1 (1969))

tilis DNA precursor pathway in order to allow SPO1 DNA synthesis to occur. The enzymes of the phage-specific pathway may be more resistant to the drug than are the host enzymes. It is also possible that SPO1 DNA, like T$_4$ DNA, may have many more replication points than host DNA during active synthesis. Finally, numerous recombination events between members of the SPO1 DNA pool in infected cells might somehow provide a release from inhibition.

BAIRD et al. (1972) compared the effect of nalidixic acid on the growth of various DNA bacteriophages by one-step growth experiments. The *E. coli* bacteriophages T$_5$,λ, T$_7$ and ϕR are strongly inhibited by the drug whereas T$_4$ and T$_2$ are only partially inhibited. The *Bacillus subtilis* bacteriophages Sp 82, Sp 5o and ϕ 29 are relatively unaffected by nalidixic acid. No correlation was found between those phages which grow in the presence of the drug and the presence of an unusual base e.g. hydroxymethyl uracil in the phage DNA. One explanation for the growth of DNA phages in the presence of nalidixic acid is discussed by the authors namely, that the resistant phage replace a drug sensitive replication function - as yet unknown - with a phage coded resistant one.

In their discussion on the effect of nalidixic acid on the replication of DNA phage M-13 SCHNECK et al. (1973) stress an important point: nalidixic acid has been found to prevent episome transfer when added to matings between F'lac (Nal^S) and F$^-$ (Nal^r). On the other hand the drug had no inhibitory effect on episomal transfer from F'lac (Nal^r) donors into F$^-$ (Nal^S) recipients (BARBOUR, 1967; HANE, 1971). It has been suggested, that DNA transfer requires DNA synthesis in the donor but not in the recipient. According to the current concept DNA synthesis is required for gene transfer in both the donor and the recipient. The transfer replication is probably an assymmetric process of DNA synthesis in which a single strand is displaced from a "rolling circle" intermediate in the donor and transferred to the recipient where a complementary strand is synthesized (BONHOEFFER and VIELMETTER, 1968; OHKI and TOMIZAWA, 1968).

SCHNECK and coworkers concluded with respect to this model, that nalidixic acid does not inhibit the conversion of single stranded DNA to the double stranded form. This conclusion was confirmed by the elegant experiments of these investigators. The effect of nalidixic acid on various stages of M_{13} replication was studied. Evidence is presented showing that the conversion of single stranded phage DNA to the double stranded replicative form is not inhibited by nalidixic acid.

Nalidixic acid has been used by PUGA and TESSMAN (1973) as a tool to study the mechanism of transcription of the mRNA of phage S_{13}. The drug was added to the phage host system to inhibit the incorporation of uridine into DNA after conversion to thymidine. Strikingly it was observed that the transcription of the phage was inhibited although the synthesis of the host (E. coli) RNA was unaffected. A theory is presented explaining that the inhibition is caused by a metabolic product of nalidixic acid that may act by intercalation into supercoiled DNA. Other explanations for the inhibition of viral mRNA synthesis by nalidixic acid are conceivable. Any explanation, however, has to include the assumption that there are some features common to both, transcription and replication in the viral system that are not found in the uninfected cell.

MOUNOLOU and PERRODIN (1968) studied and compared the effect of DNA inhibitors on the synthesis of macromolecules in Saccharomyces cerevisiae. Nalidixic acid reduces the rate of incorporation of ^{3}H-thymine into DNA in yeast, with no appreciable reduction in the rate of incorporation of labeled precursors into protein and RNA. Moreover these authors found that nalidixic acid prevents respiratory adaptation in yeast. WEHR et al. (197o) found that, in defined medium, nalidixic acid caused a temporary lag of net increase in total DNA with a corresponding inhibition in RNA synthesis. Such a lag in DNA increase could result either from a temporary inhibition of synthesis of total cell DNA, or from inhibition of only a fraction of the cellular DNA (e.g. mitochondrial DNA) accompanied by a degradation of this component.

Strikingly, in optimal growth medium nalidixic acid did not affect the rate of DNA synthesis, but caused a substantial reduction in the rate of adenine incorporation into RNA. Studies on the mode of action of nalidixic acid in *E. coli* by BOYLE et al. (1969) have also shown no noticeable effect of nalidixic acid on the rate of DNA synthesis, but a substantial reduction in the rate of adenine incorporation into RNA. It was speculated that in *E. coli*, nalidixic acid may act by affecting the regulation of the intracellular precursor pools. If this were the case, it is possible that sensitivity of saccharomyces to the drug would depend on the size or composition of the precursor pools and the consequent growth conditions of culture.

From these results it is evident that like mitomycin, nalidixic acid does not exhibit a single mode of action, but that the inhibitory effects observed depend on the physiological state of the organisms and on the conditions of growth.

b) Mutagenicity

COOK et al. (1966b) studied the mutagenic effect of nalidixic acid using streptomycin-dependence as a genetic marker. Only in proliferating cells does nalidixic acid enhance the reversion to streptomycin-independence, which is suggestive of a mutagenic action of nalidixic acid upon proliferating cultures of susceptible bacteria. Although the mechanism by which nalidixic acid induces mutation is not yet known, one possibility is discussed: A mutational event might be connected with miscoding during resumption of DNA synthesis and repair of DNA damage after removal of the drug.

c) Recombination

When nalidixic acid is added to conjugating bacteria at any time during mating, it stops genetic transfer - provided the donor bacterium is sensitive to the drug (BOUCK and ADELBERG, 197o). Strikingly, when the inhibitor is removed, transfer does not resume at the point on the chromosome where it was stopped, but begins again at the origin of transfer. By using different doses of nalidixic acid it was shown that several "hits" are necessary to inhibit recombination for early markers. The number of required hits decreases as the distance of the marker from the transfer origin increases. The transfer of lambda prophage is most sensitive to nalidixic acid when the donor bacterium is sensitive to the drug. Transfer between drug-resistant cells in the presence of nalidixic acid is inhibited after the first 15 min of mating, so that markers are progressively excluded from recombinants the later they are transferred during the cross. It was suggested that continuous DNA synthesis is necessary throughout chromosome transfer. This favors the hypothesis that transfer is mediated by DNA replication. However, the possibility cannot be ruled out that transfer and replication are separate processes, which share an identical sensitivity to nalidixic acid. Further genetic studies revealed that the difference in sensitivity of, for example, *E. coli* K-12 to the inhibitor could be located on the *E. coli* genetic map, so that the dominant relationship between resistance and sensitivity could be determined (HANE and WOOD, 1969).

d) Hemolysin Production

Nalidixic acid-resistant strains of *E. coli* K-12 were found to hemolyze equine red blood cells (WALTON and SMITH, 1968). Nalidixic acid added to the medium enhanced the hemolytic effect. The hemolysin appears to differ from that produced by bacteria possessing the transmissible hemolysin factor, and also from the two hemolysins which previously were reported to be elaborated by wild-type hemolytic strains of *E. coli*. The hemolytic effect was not confined to *E. coli* K-12, but was also exhibited by all other pathogenic and nonpathogenic nalidixic acid-resistant strains of *E. coli* examined. None of several nalidixic acid-resistant salmonella strains showed this hemolytic effect. Preliminary transduction experiments suggest that the hemolysin locus and the nalidixic acid-resistance locus are closely linked. These findings are also consistent with a possible mutagenic effect of nalidixic acid. It was shown that it was relatively easy to isolate spontaneous nalidixic acid-resistant mutants, *in vitro*, from sensitive populations of *E. coli* and salmonellas. The results of these experiments suggest that mutation to nalidixic acid resistance in *E. coli* often involves the production of a hemolysin, and that nalidixic acid present in the medium acts either by increasing the production of the hemolysin or by facilitating its release from the cell, or by a combination of these two effects.

II. Inhibitors of RNA Synthesis that Interact with the DNA Template

Antibiotics interfering with the transcription of DNA can at present
be subdivided into two classes: 1. those interacting with the DNA
template or 2. those which specifically affect either prokaryotic
or eukaryotic RNA polymerases. Chapter II is concerned with the inhibitors of the DNA template; Chapter III deals with the specific
inhibitors of prokaryotic and eukaryotic RNA polymerase.

DNA transcription *in vitro* involves several sequential processes:
1. association of the DNA template with the enzyme;
2. stabilization of the complex of the enzyme and DNA by the nucleotide which forms the 5'-terminus of the RNA chain;
3. initiation of the RNA chain by formation of the first 5'3'-phospho-
   diester linkage;
4. chain elongation;
5. chain termination, involving liberation of the synthesized polyribonucleotide chain from the template.

More simply, transcription may be considered as a three-step process,
namely: initiation, propagation and termination. The inhibitors interacting with the DNA template may affect any or all of these steps. It
now seems clear that actinomycin inhibits the propagation of RNA
chains whereas, for example, distamycin inhibits the binding of RNA
polymerase to DNA. However, no systematic studies of the transcription process have been carried out, under identical conditions with
all of these inhibitors. Hence, it is not yet possible to differen-
tiate these antibiotics according to their effect on the different
steps in RNA synthesis *in vitro*.

Thus, the inhibitors of RNA transcription have been classified according to their structural relationship and the physicochemical porper-
ties of their complexes with DNA. In accordance with this characterization, the actinomycins, the anthracyclines, the chromomycin-like
antibiotics, kanchanomycin, the peptide antibiotic distamycin, and
anthramycin are discussed separately. Subtile changes introduced by
these antibiotics into the geometry of the DNA molecule seem to vary
considerably. These observations may give us an initial impression
of the immense potentialities of conformational changes in the DNA
molecule which may occur upon binding of repressors, polymerases,
initiation and termination factors, during replication or transcription.

A. Actinomycin

1. Origin, Biological and Chemical Properties

Actinomycins, discovered in 1940 by WAKSMAN and WOODRUFF and isolated
by ROBINSON and WAKSMAN (1942), were found to be active against a va-

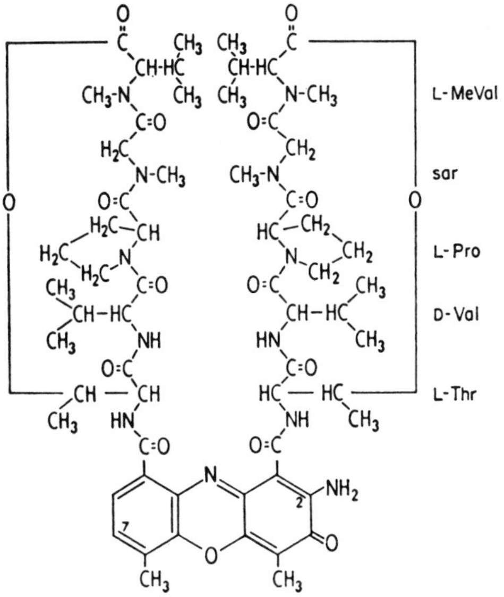

Fig. 15. Chemical structure of actinomycin C_1 (D). Abbreviations: *MeVal*, methyl-valine; *sar*, sarcosine; *Pro*, proline; *Val*, valine; *Thr*, threonine. (From: J. Mol. Biol. **68**, 2 (1972))

riety of gram-positive bacteria (FOLEY et al., 1958). This group of antibiotics inhibits the growth of experimental tumors in rats and mice (HACKMANN, 1952, 1953, 1955; OETTEL and WILHELM, 1957; FIELD et al., 1954/55), and the growth of HeLa cells in culture (NITTA, 1957; GOLDSTEIN et al., 1959).

The actinomycin antibiotics include about 3o known chromopeptides, all of which, as shown by the investigations of BROCKMANN (196o a, b) and his co-workers, are structurally related to actinomycin C_1 (= D = X_1 = I_1). The structure of actinomycin C_1 (Fig. 15) has been confirmed by several total syntheses (BROCKMANN and LACKNER, 1964 a, b, 1967). All actinomycins have the characteristic phenoxazone-ring system and the two peptide side chains which are absolutely required for biological activity. In the naturally occurring actinomycins, one or several amino acids in the cyclic peptide side chains are exchanged. The differences found among naturally produced actinomycins have been restricted to the peptide side chain. The variations have concerned the structure but never the number or configuration of the α-carbon atoms of the constituent amino acids. A survey of the nomenclature of actinomycins was made by BROCKMANN (196o b).

A number of synthetic derivatives have been prepared from actinomycins containing altered chromophores: 1) exchange of the 2-amino group for a halogen or hydroxyl (BROCKMANN, 196o a, b); 2) substitution of the hydrogen at the 7-position by -Br, NO_2, -OH, $-OCH_3$, $-NH_2$, or -NH-CO-R (BROCKMANN et al., 1966 a, b); and 3) alkylation of the 2-amino group (BROCKMANN, 196o a). In addition, actomycins containing side chains with different amino acids have been synthesized (BROCKMANN and LACKNER, 1964 b), including two actinomycins with altered substituents at the 4 and 6 positions (BROCKMANN and SEELA, 1965). Two

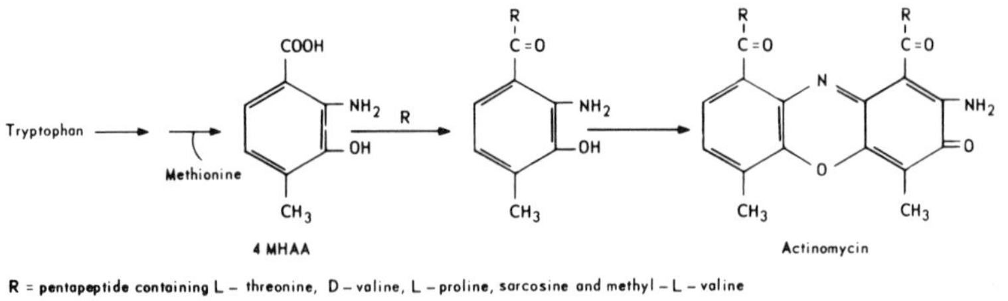

R = pentapeptide containing L – threonine, D – valine, L – proline, sarcosine and methyl – L – valine

Fig. 16. Proposed pathway for actinomycin biosynthesis. (According to WEISSBACH et al., 1965. From: Biochem. Biophys. Res. Commun. <u>19</u>, 525 (1965))

semi-actinomycins lacking one peptide ring and two optical antipodes of actinomycin C_1 (BROCKMANN and SCHRAMM, 1966) have also been synthesized.

The chromophore actinocin can be formed enzymatically (WEISSBACH and KATZ, 1961; KATZ and WEISSBACH, 1962) by the condensation of 2 moles of 4-methyl-3-hydroxy-anthranilic acid. The phenoxazinone-condensing enzyme will also form phenoxazinones from anthranyloyl peptides. A mechanism for the synthesis of the antibiotic, formulated by WEISSBACH et al. (1965) is shown in Figure 16. A pentapeptide is added to 4-methyl-3-hydroxy-anthranilic acid. Two of these molecules then condense to yield the antibiotic. 4-Methyl-3-hydroxy-anthranilic acid is an intermediate in the conversion of dl-tryptophan to actinocin.

By supplementing the culture medium of actinomycin-producing strains of Streptomyces with a variety of amino acids, it has been possible to influence the proportions of the different actinomycins synthesized (SCHMIDT-KASTNER, 1956; KATZ, 1960).

2. On the Molecular Mechanism of Action

a) Interaction with DNA, Poly- and Mononucleotides *in Vitro*

Interaction between actinomycin C_1 and double helical DNA has been demonstrated by cosedimentation of actinomycin with DNA in the analytical ultracentrifuge and other methods by KERSTEN et al. (1960), KIRK (1960), KAWAMATA and IMANISHI (1960, 1961) and KERSTEN (1961 a).

The reaction is accompanied by a change in the spectral properties of the chromophore: The absorption maximum around 440 mµ shifts by about 12 nm to longer wavelengths with a 30% decrease in the extinction coefficient. Similar spectral changes in the absorption spectrum of actinomycin were found in the presence of purine and purine ribonucleosides, indicating that actinomycin also interacts with nucleic acid constituents (KERSTEN, 1961 b), preferentially with deoxyguanosine (Fig. 17). From the observation of GOLDBERG et al. (1962) that a guanine-free DNA, for example dAT, does not bind actinomycin, it was concluded that in the binding of actinomycin to DNA, deoxyguanosine residues play an important role - a point which was discussed in more detail by PULLMAN (1964).

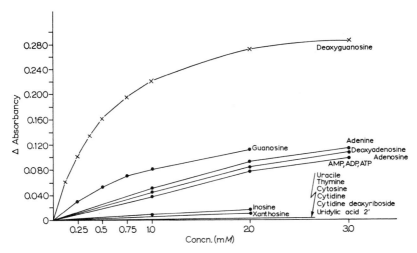

Fig. 17. Differences between maximum and minimum of the difference spectrum, measured upon interaction of actinomycin with nucleosides and plotted against the concentration of the nucleosides. (According to KERSTEN, 1961 b)

WELLS (1969) and WELLS and LARSON (1970) have studied the binding of actinomycin to 17 different synthetic DNA polymers by a variety of techniques including equilibrium dialysis, *in vitro* transcription and analytical buoyant density centrifugation. The results of these experiments can be briefly summarized: the presence of deoxyguanosine is not necessary for the binding of actinomycin to all kinds of DNA, since poly dI binds actinomycin almost as tightly as does native DNA; the presence of deoxyguanosine is not sufficient for binding since poly d(A-T-C) · poly d(G-A-T) which contains 33% G+C binds virtually no actinomycin. A marked nucleotide sequence preference exists for the binding reaction. By comparing the binding capacities of sequence isomeric DNAs, the isomer which contains both purines and pyrimidines on both complementary strands, binds more actinomycin and binds it more tightly than the isomer containing all the purines in one strand and all the pyrimidines in the complementary strand. These findings strongly suggest that the apparent specificity of guanine actinomycin binding to DNA must be a secondary effect and caused by an influence of this base on a certain steric and electronic environment. Most important for actinomycin association with DNA is the presence of a suitable DNA configuration. The results are not consistant with the model of HAMILTON et al. (1963), but favor the hydrogen-bonded intercalation model of MÜLLER and CROTHERS (1968; see sections b and c).

As a consequence of complex formation with DNA, actinomycin inhibits the DNA-dependent RNA polymerase, whereas DNA-dependent synthesis of DNA with DNA polymerase I enzyme *in vitro* is only slightly affected (HARBERS and MÜLLER, 1962; GOLDBERG et al., 1962; HURWITZ et al., 1962; HARTMANN and COY, 1962). These results suggested that the biological basis of actinomycin action depends on its association with DNA within the cell, thereby inhibiting DNA-dependent processes. Surveys on this topic were presented by REICH and GOLDBERG (1964), REICH et al. (1967), and GOLDBERG and FRIEDMAN (1971). Since then, several groups have studied the detailed mechanism of the interaction of actinomycin with DNA, because the chromopeptide antibiotic and its association with DNA might serve as a primitive model for interaction of repressors with DNA (KERSTEN et al., 1966).

b) Interaction with Deoxyguanosine

Spectral changes produced in actinomycin solutions have been found upon addition of several purine derivatives in the following decreasing order of effectiveness: deoxyguanosine>guanosine>adenine = AMP = = ATP = adenosine = deoxyadenosine. Inosine, xanthosine and normal pyrimidine bases or their nucleosides were inactive (KERSTEN, 1961 b; KERSTEN and KERSTEN, 1962 b). Interaction of actinomycin with deoxyguanosine or with 5-deoxyguanylic acid was studied by MÜLLER and SPATZ (1965), ARISON and HOOGSTEEN (1970), and SOBELL et al. (1971 a, b).

MÜLLER and SPATZ determined the molecular weight of the complex by sedimentation analysis and studied stoichiometry and kinetics of the interaction by using the temperature jump method. It was concluded that the complex between actinomycin C_3 and deoxyguanosine contained two moles of actinomycin and two moles of deoxyguanosine. The stability constants of the nucleoside complexes from 22 different natural or partially synthetic actinomycins were determined. A structure of the complex is proposed in which a hydrogen bridge is formed, between the quinone oxygen of the actinomycins and the amino group of deoxyguanosine. Besides this, hydrophobic interactions between deoxyribose of the nucleoside and certain alkyl groups of the peptide chains of the actinomycin stabilized the complex.

In a preceding paper MÜLLER and EMME (1965) showed that in aqueous solution actinomycins dimerize at a concentration of 1×10^{-5} M and form oligomers at concentrations higher than 10^{-3} M. Two different structures are proposed for the dimers of actinomycin in which the peptide rings fit into each other, as shown in Fig. 18. The two chromophores can be oriented differently, so that in one case a symmetric, and in the other an asymmetric structure arises. The hydrophobic interactions between the alkyl residues of the amino acids stabilize the dimer.

The stoichiometry of the actinomycin-deoxyguanosine complex indicates that upon interaction, the actinomycin dimer is not dissociated into monomers. A complex is postulated between an actinomycin dimer and two deoxyguanosine molecules. Substitutions at the actinomycins which increase the redox potential of the quinone component decrease the stability of the complex. The quinone oxygen is therefore suggested to be involved in the formation of a hydrogen bond. The H-donor is assumed to be the amino group of the guanine residues. Reduction of the quinone oxygen in actinomycin leads to a derivative which is no longer able to bind to deoxyguanosine. Pretreatment of deoxyguanosine with formaldehyde also inhibits complex formation with actinomycin, indicating that either the amino group in position 2 or the amino group in position 1 or the purine functions as donor of the hydrogen. Stereochemical considerations favor hydrogen bonding by the amino group.

A detailed 220-MHz proton magnetic resonance study was made on the dependence of concentration, pD, salt and temperature of the actinomycin D spectrum in D_2O. The results confirm that actinomycin D aggregates to form a dimer at the concentration ranges and temperatures covered in the work of ANGERMAN et al. (1972). These authors conclude that the dimer is formed by an interaction between the actinocyl chro-

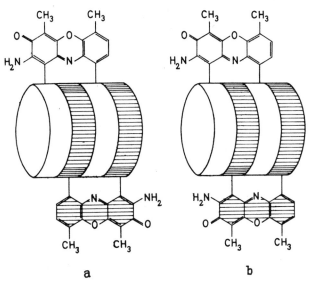

a   b

Fig. 18. The two alternatives of the actinomycin dimer. (According to MÜLLER and SPATZ, 1965. From: Z. Naturforsch. 2ob, 849 (1965))

mophore groups only. Based upon the direction and relative magnitudes of the shift trends for the actinocyl group protons, they further concluded that the actinocyl groups stack vertically in the dimer with one chromophore inverted with respect to the other (Fig. 19). An interpretation of the dimer structure and the resultant shift trends is given in terms of the diamagnetic shielding anisotropy of the actinocyl chromophore group. The chemical shift versus concentration curves for the actinocyl signals were analyzed. Dimerization equilibrium constants of $2.7 \times 10^3 \ M^{-1}$ and $1.4 \times 10^3 \ M^{-1}$ at $18^\circ C$ (pD = 7.2) were obtained.

Variation in the composition of the peptide rings also influences the stability of the deoxyguanosine-actinomycin complex. If threonine is substituted by serine, and thus the threonine-methyl group is missing, a less stable complex results. It is therefore suggested that alkyl groups stabilize the complex.

Nuclear magnetic resonance (NMR) spectral studies on actinomycin D and on the complex with 5'-deoxyguanylic acid were reported by ARISON and HOOGSTEEN (197o). NMR spectra of actinomycin D in various solvent systems showed that the molecule has a relatively high degree of molecular symmetry, which is best achieved by orienting the pentapeptides so that their planes are parallel to each other and perpendicular to that of the phenoxazone nucleus. Such an arrangement would also be consistent with a hydrogen bonding scheme suggested for the thr-NH protons, since this could fix the pentapeptides in the proposed configuration. The location of the threonine methyls above the plane of the phenoxazone ring is deduced from the study of the complex with deoxyguanylic acid. From the NMR data it is evident that on complex formation all the protons on the chromophore were shifted upfield, while both threonyl-methyl groups were deshielded. These displacements require a position of the base above the chromophore, so

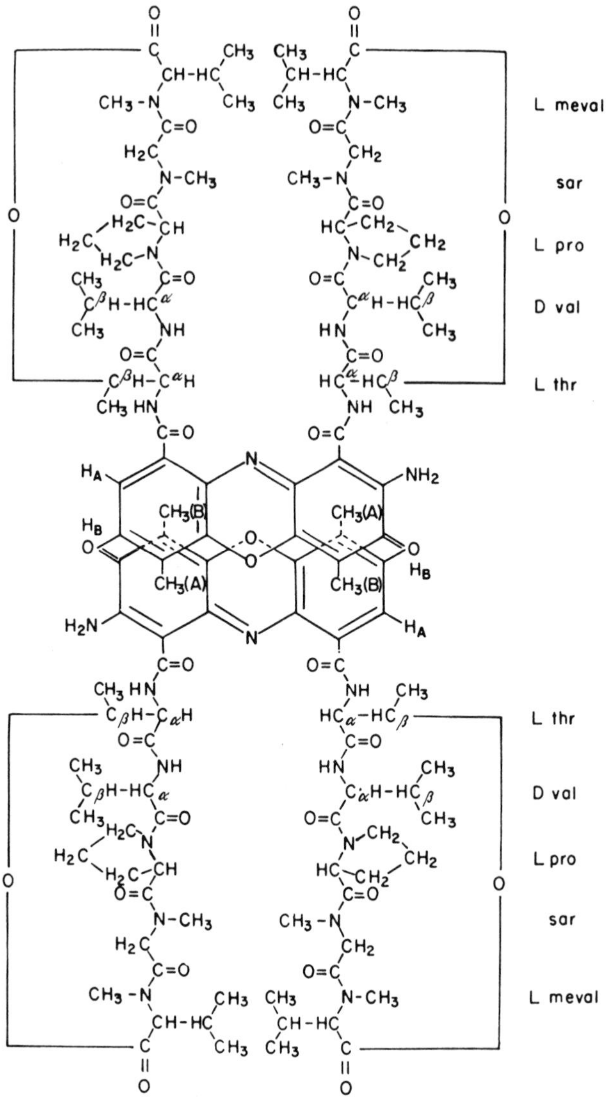

Fig. 19. Representation of the actinomycin D dimer structure. (According to ANGERMAN et al., 1972. From: Biochemistry 11, 241o (1972))

that the planes of the two systems are oriented parallel to one another. The NMR data clearly favor a base-stacking model for the complex. This picture of the complex is a qualitative one. Measurements at ratios of dGMP to actinomycin of less than 2:1 could not be investigated owing to the precipitation of actinomycin.

The three-dimensional structure of a crystalline complex containing actinomycin D and deoxyguanosine was evaluated by X-ray analysis (SOBELL et al., 1971 a). Ethanol-water solutions (5o%) of actinomycin

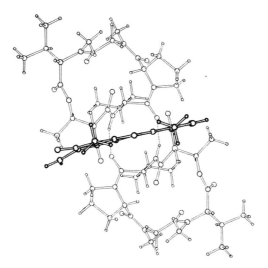

Fig. 2o. Stereochemical model of actinomycin. (According to SOBELL and JAIN, 1972. From: J. Mol. Biol. **68**, 26 (1972))

D and deoxyguanosine (mixed 1:1 and 1:2) were prepared and evaporated. Large single crystals developed after a few days, and ultraviolet absorption spectra of solutions obtained from washed single crystals indicated a 1:2 stoichiometric complex between actinomycin and deoxyguanosine. The stoichiometry of the solid-state complex does not reflect the stoichiometry of the complex in solution. Two crystalline modifications of this complex have been isolated and the crystal data are presented. Refinement and further structural details of the actinomycin-deoxyguanosine crystalline complex were published by JAIN and SOBELL (1972). The earlier postulated twofold symmetry of actinomycin reflects an approximate dyad axis lying roughly along a vector connecting the O-N bridging atoms in the phenoxazone ring. Although this symmetry is not exact, both polypeptide chains closely conform to this non-crystallographic twofold symmetry. The conformations of the peptide linkages are as follows: L-threonine-D-valine, trans; D-valine-L-proline, cis; L-proline-sarcosine, cis; sarcosine-L-methylvaline, trans; L-threonine-carboxamide carbonyl oxygen and carbon of chromophore, trans. A strong hydrogen bond exists between neighboring cyclic pentapeptide chains connecting the N-H of one D-valine residue with the carbonyl oxygen of the other D-valine residues. No other hydrogen bonds stabilize the actinomycin structure either between chains or within chains. These findings are consistent with NMR data implicating the D-valine N-H groups as hydrogen bond donors. The quinoidal portion of the phenoxazone ring system is slightly twisted in a propeller-like fashion with respect to the aromatic portion of the chromophore (Fig. 2o).

The complete crystal structure of the actinomycin-deoxyguanosine complex is presented. X-ray analysis was performed by using 7-bromoactinomycin C_1. This compound co-crystallizes with deoxyguanosine to form an isomorphous structure which has almost identical unit-cell parameters, with one crystal modification of the complex between actinomycin C_1 and deoxyguanosine (light-atom complex). Crystal data were

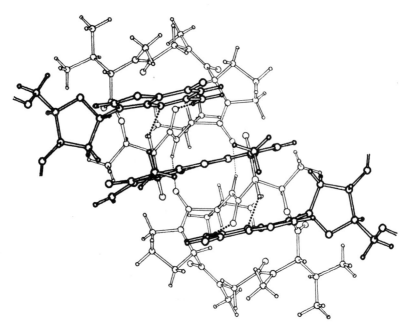

Fig. 21. Stereochemical model of the actinomycin-deoxyguanosine complex. (According to SOBELL and JAIN, 1972. From: J. Mol. Biol. **68**,
26 (1972)

collected from both the light-atom and heavy-atom complexes at room
temperature.

Like actinomycin the complex between deoxyguanosine and actinomycin
has a twofold symmetry. The two deoxyguanosine molecules interact with
the two cyclic pentapeptide residues and stack on alternate sides of
the phenoxazone ring system. A strong hydrogen bond connects the gua-
nine-2-amino group with the carbonyl oxygen of L-threonine residue.
A weaker hydrogen bond connects the guanine N_3 ring nitrogen with the
N-H group of this same L-threonine residue. The conformation of both
deoxyguanosine molecules is anti. The sugar residues of both deoxyguanosine molecules are in close steric juxtaposition with the iso-
propyl groups of the L-methyl-valine residues. Hydrophobic contacts
as well as the stacking of guanine and phenoxazone rings provide stabilization to the complex (Fig.21).

The complete crystal structure is shown in Figure 22. The strucutre
contains 14o atoms in the asymmetric unit; one actinomycin, two deoxyguanosine and 12 water molecules. The structure is stabilized by
a total of 152 hydrogen bonds in each unit cell. Approximately 8o%
of these involve interactions with water structure. The water structure appears to be well ordered and rigidly held in this structure
(Fig. 22).

The actinomycin-deoxyguanosine complex differs in a fundamental way
from base-paired structures in which purine-pyrimidine hydrogen bonding is the mode of association between guanine and phenoxazone rings.
In addition to stacking interactions and hydrophobic bonding, the ac-

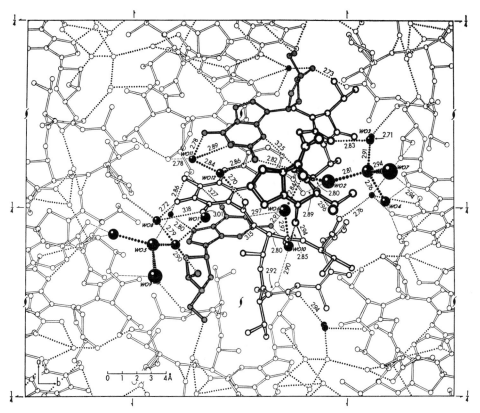

Fig. 22. The actinomycin-deoxyguanosine crystal structure as viewed down the c axis. Black circles with highlights indicate water molecules. Deoxyguanosine molecules have been shaded with striped lines, and are seen stacked above and below the plane of the chromophore residue. The actinomycin molecule is indicated with dark open lines and circles. The surrounding crystal structure is shown with lighter lines. Black dots indicate hydrogen bonds between atoms at neighboring levels. Dashed lines indicate hydrogen bonds between atoms located at different levels in adjoining unit cells. Additional hydrogen bonding in the surrounding structure is indicated by small open circles. (According to JAIN and SOBELL, 1972. From: J. Mol. Biol. <u>68</u>, 16 (1972))

tinomycin-deoxyguanosine complex demonstrates hydrogen bonding between the peptide portion of the actinomycin molecule and the guanine ring. This provides additional stabilization of the complex and explains the added specificity which deoxyguanosine exhibits.

Actinomycin D mononucleotid interactions as studied with proton magnetic resonance by KRUGH and NEELY (1973 a, b) lead to the following conclusions: actinomycin D has two binding sites for deoxyguanosine 5'-monophosphate and guanosine 5'-monophosphate. The adenine nucleotides bind to the same binding sites. The complex consists of stacking one of the guanine bases on each side of the actinomycin D chro-

mophore. The geometry of the actinomycin D - deoxyguanosine 5'-mono-
phosphate complex is similar to that of the complex with deoxyguanosine (SOBELL et al., 1971). The complexes of actinomycin with other
nucleotides have similar geometry. Deoxyguanosine 5'-monophosphate
appears to have a much larger binding constant than guanosine 5'-mo-
nophosphate for the binding site on the quinoid portion of the chromophore.

Sequence specificty of actinomycin binding to oligonucleotides has
been investigated by SCHARA and MÜLLER (1972). G-containing dinucleotides of the type d (pX-G), d (pG-X), and d (pG-C) and a tetranucleo-
tide of the type d (pC-T-A-G) were used. The dinucleotide d (pG-C)
binds actinomycin in a 2:1 complex (two dinucleotide per dye molecules) of higher stability than the other dinucleotides and the tetra-
nucleotide. These data agree well with the X-ray data of the actino-
mycin-dG crystal (SOBELL et al., 1971) showing that the actinomycin
chromophore is linked to the 3'-side of the d-G-residue. For review
see SOBELL (1973).

c) Proposed Models for the Actinomycin-DNA Complex

Three different models for the structure of the complex between actinomycin and DNA are described.

The first model was proposed by HAMILTON et al. (1963; Fig. 23) and is
based on X-ray diffraction and molecular model-building studies. According to this model, actinomycin is considered to be located in
the minor groove of helical DNA, with which it can form up to seven
hydrogen bonds. The geometry of three of these H-bonds has been studied in detail and has been found to be stereochemically satisfactory.
The authors argue that the properties of the complex deduced from this
model fit most of the known facts concerning the reaction of actinomycin with DNA and the associated inhibition of DNA-dependent RNA
synthesis. According to this model, the quinoidal oxygen of actinomycin acts as an H-bond acceptor, H is donated by the amino group in
the C_2 position of guanine. Another hydrogen bridge is formed between
the N-3 of guanine and the amino group of the actinomycin molecule
in position 3. The ring oxygen of the deoxyribose also functions as
a hydrogen bond acceptor, accepting hydrogen bonds also from the amino
group on the actinomycin chromophore. The arguments are: Alterations
involved in the amino group of actinomycin would eliminate one or
both of the hydrogen bonds formed; reduction of the quinoidal oxygen
would also restrict the ability of actinomycin to bind to DNA; the
lactone rings stabilize the cyclic peptide side chains in a conformation permitting the formation of four additional H-bonds between
four peptide-NH groups and phosphodiester oxygens of the DNA strand
opposite that of guanine interacting with the chromophore. The model
depends critically on the relative positions of the DNA constituents
as they are disposed in helical DNA in the B conformation. This is in
accord with the poor binding of actinomycin, if at all, to single-
stranded DNA and absence of binding to RNAs such as reovirus RNA
and sRNA (SHATKIN, 1965).

The second model for the actinomycin-DNA complex was proposed by MÜLLER
and CROTHERS (1968): To study the structure of the DNA-actinomycin
complex in solution, they used, as with the deoxyguanosine-actinomycin complex, equilibrium, hydrodynamic and kinetic measurements.

Equilibrium measurements revealed that, very probably, a large part
of the specificity of actinomycin binding to DNA lies in the mode of

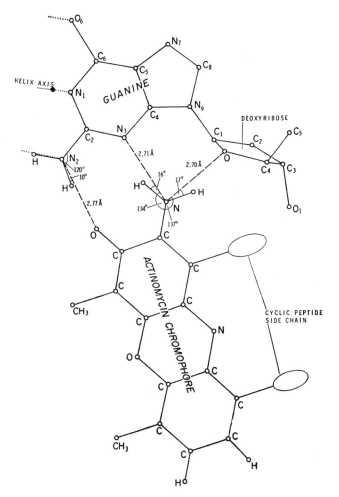

Fig. 23. Model of actinomycin-DNA interaction. (According to HAMILTON et al., 1963. From: Nature <u>198</u>, 539 (1963))

binding of the chromophore and possibly of the carboxamide group, and that the peptide rings are certainly not the sole determinants of the specificity. Though actinomycin binding shows preference for the GC pairs, the number of binding sites was found to be much fewer than the total number of GC pairs. The equilibrium data suggest that the apparent binding constant of several derivatives of actinomycin are related to the presence of the carboxamide NH. A structural distortion of the DNA helix upon complex formation is probably generated by hydrogen bonding to this group. It is concluded that the structural distortion of the helix prevents binding of actinomycin at nearby GC pairs. With the assumption that the GC pairs are randomly distributed along the polymer, it was possible to calculate binding isotherms for DNA of any given GC content (CROTHERS et al., 1968).

GELLERT et al. (1965) reported that the viscosity of DNA decreased when actinomycin was added to a solution of DNA. The hydrodynamic properties of DNA in the presence of actinomycin were found to be concentration-dependent; at low concentrations a decrease and at high concentrations an increase in viscosity was observed (KERSTEN et al., 1966). MÜLLER and CROTHERS confirmed that either an increase or a decrease in viscosity could occur and reported that two variables proved to be important in this analysis: the extent of binding and the molecular weight of the DNA sample. With DNA of high molecular weight the viscosity first decreases, whereas the intrinsic viscosity of low molecular weight DNA rises immediately upon addition of actinomycin. The rise in viscosity and decrease in the sedimentation coefficient were the first indications that actinomycin might be inserted between adjacent base pairs, like the acridines (LERMAN, 1961). MÜLLER and CROTHERS argue that the hydrodynamic changes produced in high molecular weight DNA by actinomycin are side effects due to the peptide rings, and that these mask the hydrodynamic changes expected upon intercalation of the chromophore.

The kinetic measurements, with a stopped-flow apparatus, contributed information concerning the actinomycin-DNA complex which was not available from either equilibrium or hydrodynamic studies. The kinetic measurements indicated that a multiplicity of complex forms may exist, not only as kinetic intermediates, but also in the final equilibrium state. The requirement for at least five time constants to characterize the association reaction between actinomycin C and DNA indicates that there are at least five forms of the complex. This complicated behavior is different from the simple kinetics observed by MÜLLER and SPATZ (1965) for the formation of a complex between actinomycin and deoxyguanosine. Actinomine, lacking the complicated cyclic peptide rings (Fig. 24), but still containing the carboxamide arrangement in the side chain, reacts much faster, and more simply, with DNA. This indicates that the peptide rings may undergo conformational changes. The slow character of the actinomycin-DNA interaction is thought to be a consequence of the sterically restricted structure of the cyclic peptide rings. The five complex forms are formulated AD_1, AD_2, AD_3, AD_4, AD_5 (Fig. 25), in which AD_1 and AD_2 are interpreted as kinetic intermediates and AD_3, AD_4 and AD_5 as structures in which both peptide rings (AD_5) or one of the two rings (AD_3 or AD_4) have undergone conformational changes.

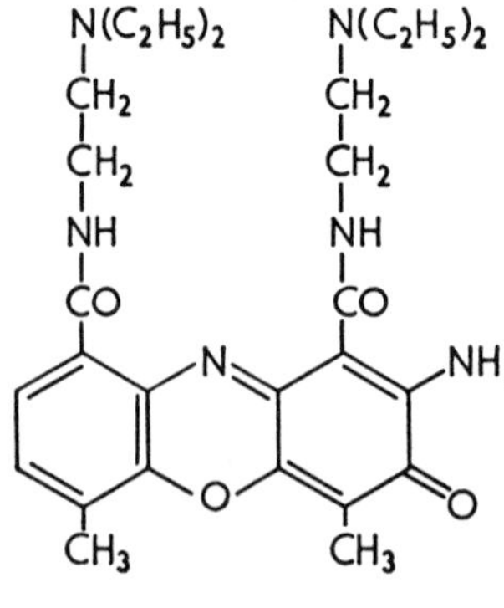

Fig. 24. Structure of actinomine. (From: J. Mol. Biol. 35, 255 (1968))

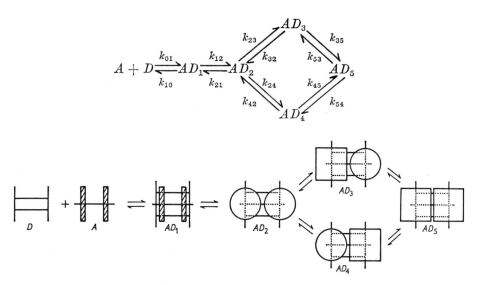

Fig. 25. Schematic drawing of the proposed mechanism for the reaction of actinomycin with DNA. (According to MÜLLER and CROTHERS, 1968. From: J. Mol. Biol. <u>35</u>, 280 and 286 (1968))

Substitution of the actinomycin chromophore in position 7 considerably decreases the rate of reaction between actinomycin and DNA. The second-order rate constant was found to be reduced by three orders of magnitude when the synthetic 7-N-trimethyl-acetyl derivative of actinomycin was tested. The bulky group in position 7 is assumed to make the insertion reaction difficult and to account for the dramatic decrease in the combination rate constant.

The most satisfactory model that is in agreement with the measurements, starts with the hypothesis that the chromophore is intercalated between the base pairs in the DNA structure. The discovery of π-complexes with the actinomycin chromophore makes such a mechanism of interaction with DNA plausible, and electronic interaction energy therms provide a ready explanation for the guanine specificity. The importance of the carboxamide NH group for the complex structure follows directly from an intercalation mechanism, but not from any simple model of external binding. The hydrodynamic measurements make it clear that the length of the DNA molecule is increased by actinomycin and by actinomycin analogs lacking the peptide rings. It is suggested that the chromophore is inserted from the small groove, so that the peptide rings project into the small groove. It is assumed that intercalation can occur adjacent to any GC pair and that the quinoid part of the chromophore is next to the cytosine residue. In this orientation the nitrogen of the chromophore ring falls directly under the purine 2-amino group. The apparent specificity for purine 2-NH$_2$, as proposed by CERAMI et al. (1967), is ascribed to this juxtaposition. The carboxamide-NH can then be hydrogen-bonded to the deoxyribose-ring oxygens on each side of the double helix. It is suggested that these bonds are the source of a local distortion of the helix, which prevents binding of another actinomycin closer than six base pairs away. It is speculated that hydrogen bond from the chromophore amino

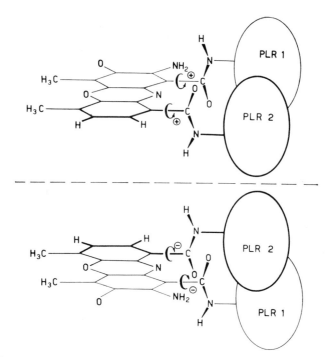

Fig. 26. Schematic drawing of the two "atropisomers" of actinomycin. PLR 1 and PLR 2 are representatives of the two peptide lactone rings. + and - signs indicate the chirality sense. (According to DE SANTIS and SAVINO, 197o. From: Nature 227, 1238 (197o))

group to the phosphate oxygen lies between the cytosine and the base on the other side of the actinomycin chromophore. The exact placement of the peptide rings is not yet clear.

That the actinomycin molecule has to undergo a conformational change to be able to associate with DNA was suggested by GURSKY (1969). Results on the optical rotatory dispersion (ORD) and circular dichroism (CD) data for the actinomycin-DNA complex and for actinomycin alone (ZIFFER et al., 1968; YAMAOKA and ZIFFER, 1968; CROTHERS et al., 1968; ASCOLI et al., 197o) indicate that the rotational strength of the optically active transitions in the visible region which can be ascribed to the chromophore of actinomycin increases in the presence of DNA. CD spectra of actinomycin in different solvents strongly indicate the existence of two opposite chiralities for the molecule in different solvents which give rise to two different conformations (shown in Fig. 26). These two conformations may be considered as two local optical antipodes which account for the close antisymmetry of the CD spectra. Two different conformations of actinomycins were also proposed by CONTI and DE SANTIS (197o) on the basis of NMR spectroscopy (Fig. 27). A model of actinomycin D was constructed which connects the two peptide lactone rings at the chromophoric group and demonstrates the possibility of two hydrogen bonds between the NH- of D-valine residues and the CO- of N-methyl valine residues. The difference between the two cyclic chains was ascribed in this case to the non-equivalence of the conformation at the L-threonine because of the different at-

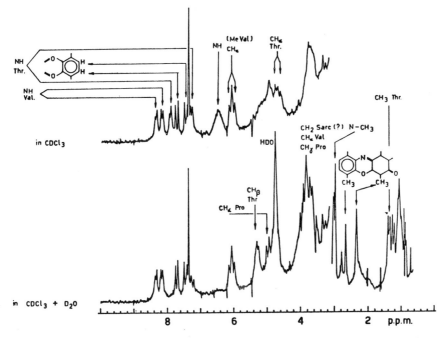

Fig. 27. NMR spectra of actinomycin D. (According to CONTI and DE SANTIS, 197o. From: Nature <u>227,</u> 1239 (197o))

tachments of the two rings to the chromophoric group. The opposite helicity of the two conformations is assumed to be very critical for the stereospecificity of the interaction with DNA.

The third actinomycin-DNA model was presented by SOBELL et al. (1971 a,b), and by SOBELL and JAIN (1972). The molecular model of the actinomycin-DNA complex is based on the geometry determined in the actinomycin-deoxyguanosine crystalline complex (Figs. 28 a, b, 29) and the double helical hexanucleotide complex with actinomycin (Fig. 3o). The model provides a structural explanation for some physical and biochemical data related to the interaction of actinomycin with DNA and correlates the three-dimensional structure of the antibiotic with its biological action.

Corey-Paulin-Koltun (CPK) space-filling models were found to be ideally suited for this purpose. For preliminary studies of the complex, a hexanucleotide sequence dApTpGpCpApT was used. It has the advantage of being self-complementary and of forming a double helical structure which has twofold symmetry.

The major structural features of this actinomycin-DNA stereochemical model are:
1. Intercalation of the phenoxazone ring system between the base-paired dinucleotide sequence, GpC. This formation results in movements of both guanine residues on opposite strands approximately 1.25 Å towards the axis of the helix and an unwinding of the helix by -18°. Both the chromophore and the G-C base pairs are tilted 1o° to a plane

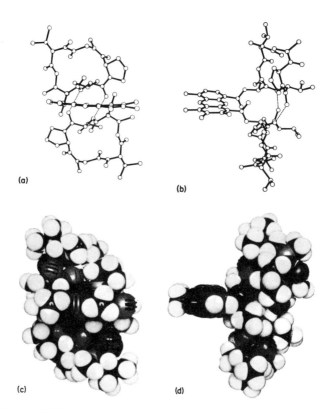

Fig. 28a. Space-filling model of the actinomycin molecule as it exists in a crystalline complex with deoxyguanosine. *a*) and *b*) Computer-
drawn illustrations of the actinomycin molecule viewed *a*) down its
approximate dyad axis and *b*) from a sidewise direction. *c*) and *d*)
Space-filling model of actinomycin viewed in the same directions.
a) and *c*) after SOBELL, H.M., et al., 1971. From: Nature New Biology <u>231</u>, 2o1, with permission; *b*) reproduced with permission from
SOBELL et al., 1971b

perpendicular to the helix's axis. The non-planarity of the phenoxazone ring system alters the stacking of guanine and cytosine residues
immediately above and below the quinoidal portion of the phenoxazone
ring system, resulting in a twist of 8° between G-C base pairs. A local asymmetry in the DNA structure at the site of intercalation results.
2. Conformational changes occur in the sugar phosphate backbone.
3. The peptide portion of the actinomycin molecule lying in the narrow
groove of the DNA helix is connected through hydrogen bonds to guanine residues. A strong hydrogen bond connects the guanine-2-amino group
with the carbonyl oxygen of the L-threonine residue, while a weaker
hydrogen bond connects the guanine N_3 ring nitrogen with the N-H group
on this same L-threonine residue. The hydrogen-bonded configuration
closely resembles that found in the actinomycin-deoxyguanosine crystalline complex. In addition to these hydrogen bonds numerous van der
Waals contacts occur between hydrophobic groups on the actinomycin
molecule and the DNA structure and these additionally stabilize the

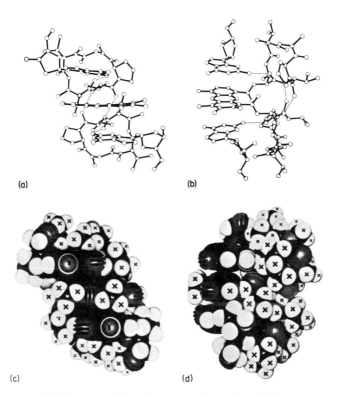

(a)   (b)

(c)   (d)

Fig. 28b. Space-filling model of a complex containing one molecule
of actinomycin and two molecules of deoxyguanosine. *a*) and *b*) Com-
puter-drawn illustrations of the complex viewed *a*) down its approximate dyad axis and *b*) from a sidewise direction. *c*) and *d*) Space-
filling model of the complex viewed in the same directions. To help
distinguish the antibiotic molecule (cf. Fig. 28a) its hydrogen
atoms have been marked with crosses. An additional hydrogen bond
(not shown) connects the N-H group of L-threonine to the N(3) ring
nitrogen on each guanine moiety. (After SOBELL et al., 1971 a, with
permission)

complex. The two cyclic pentapeptide chains are related by twofold
symmetry. The axis of symmetry relating subunits on actinomycin co-
incides with the axis of symmetry relating the sugar phosphate backbone and base sequences on the DNA helix.
4. Thymine-adenine base pairs immediately proximal to the intercalated GpC dinucleotide sequence undergo unwinding. The twist angle is
reduced from 36° in the B form of DNA to 28°. The AT base pairs fur-
thest from the intercalation site are also tilted 10° to a plane perpendicular to the axis of the helix; however they are related by a
twist of 36° from the preceding TA base pairs.
5. An additional hydrogen bond may connect the 2-amino group on actinomycin with a phosphate oxygen on the neighboring sugar phosphate
chain through a water bridge. Another possible hydrogen bond involves
the interaction of this amino group with the furanose ring oxygen on
the deoxycytidine residue.

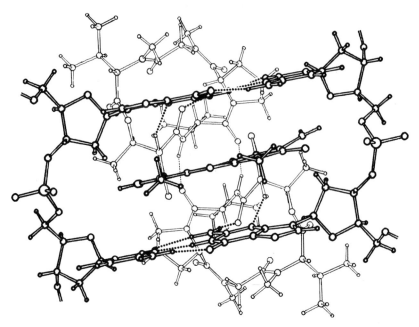

Fig. 29. Stereochemical model of the actinomycin-d-GpC complex. (According to SOBELL and JAIN, 1972. From: J. Mol. Biol. <u>68</u>, 26 (1972))

The stereochemical model for actinomycin-DNA binding is consistent with evidence concerning polynucleotide conformation, favoring intercalation. The model is a synthesis of the two previous models described above.

Further evidence favoring an intercalation model of the actinomycin DNA complex is presented by ZIPPER et al. (1972) who investigated the complex of actinomycin C_3 with small-angle X-ray scattering. The experimental data obtained are inconsistent with an outside-binding model.

SOBELL (1973) suggests that the binding of actinomycin to DNA demonstrates a general principle which several classes of proteins may utilize in recognizing symmetrically arranged nucleotide sequences on the DNA helix. This discussion is based on the findings of BERNARDI (1968) and KELLY and SMITH (1970) who observed a deoxyribonuclease from the spleen, which is a dimer containing identical subunits. BERNARDI postulated that if the subunits of this enzyme were related by twofold symmetry, this would allow the enzyme to recognize the dyad axis on DNA, and if each subunit had an active site, permit simultaneous double-strand scission of the sugar phosphate backbone. A consequence of this type of recognition would involve symmetrically arranged sequences such as CpG or GpC, the latter being the sequence which actinomycin bind preferentially. A symmetrically arranged hexanucleotide sequence is also recognized by the restriction enzyme isolated from *Hemophilus influenzae*. Although the subunit character of this enzyme has not yet been established KELLY and SMITH (1970) postulated a similar protein-nucleic acid recognition pattern for this restriction enzyme.

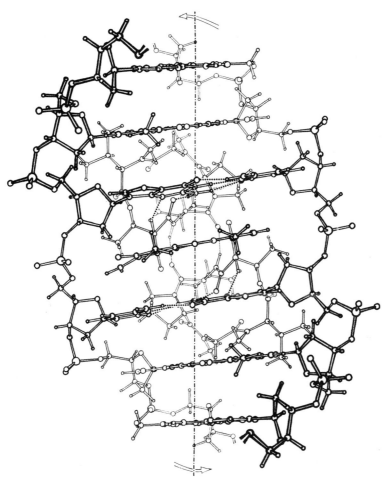

Fig. 3o. Stereochemical model of the actinomycin-d-ApTpGpCpApT complex. (According to SOBELL and JAIN, 1972. From: J. Mol. Biol. <u>68</u>, 26 (1972))

The binding of actinomycin to DNA (in particular to GpC sequences) and its specificity in inhibiting the RNA polymerase reaction, suggest a primitive "repressor operator character" for this complex, which may prove to have more general significance with regard to the recognition of naturally occurring operators by repressors. If a repressor molecule has identical subunits related by twofold symmetry when it binds to DNA, then a necessary consequence is that the base sequence in the operator has twofold symmetry. It is suggested that this general principle for dimer recognition can be extended to include tetramer recognition as well.

Although these are actually speculations which would lead to the elucidation of the very important problems of nucleic acid-protein interactions, the model of the actinomycin-DNA complex is still an idealization and that the final conformation can only be obtained using the method of single crystal X-ray analysis.

These results are not consistent with the model of HAMILTON et al. (1963), but interpretation of the binding of actinomycin to DNA is difficult because, in addition to double helical DNA, single-stranded fd-phage DNA and apyrimidinic acid also form complexes with actino- mycin which are similar in composition to that formed with native DNA (LIERSCH and HARTMANN, 1965). Furthermore actinomycin interacts with denatured DNA (KERSTEN, 1961a). These findings might be explained by the observations of CAVALIERI and NEMCHIN (1964, 1968) who found by equilibrium dialysis and light scattering two types of binding sites, one of which had binding constants about 5o times greater than the other. One type of binding might therefore be involved in the inter- action with single-stranded or denatured DNA. GELLERT et al. (1965), by measuring the thermodynamics of actinomycin interaction with DNA, also describe two types of binding sites. In heat denaturation of the DNA-actinomycin complex these authors did not observe considerable amounts of actinomycin bound to DNA under the experimental conditions used, and concluded that actinomycin did not associate with denatured DNA. But this is only valid for high temperatures at which also de- oxyguanosine does not interact with the antibiotic (KERSTEN and KERSTEN, 1962 b).

The association of actinomycin D and deoxyribonucleotides as a model for the binding of the drug to DNA has been investigated spectro- photometrically. The change in the absorbance of actinomycin D at 425 nm is plotted against added nucleotide concentration for a series of nucleotides. The nucleotides pdC (5'-dCMP), pdC-dC, and pdT-dC over a broad concentration range did not affect the spectra of acti- nomycin D. The dinucleotide pdA-dC had a binding curve similar to the pdA-dT curve. The association with pdG-dC is much stronger than with the other dinucleotides containing guanine. However, all nucleotides containing guanine will form a complex with actinomycin D and all nu- cleotides with the general structure pdN-dG exhibit virtually identi- cal binding curves. The dinucleotides pdG-dA and pdG-dT exhibit bind- ing curves that are quite different from the pdN-dG curves and the pdG-dC curve. From this result KRUGH (1972) concluded that actinomy- cin D had a great deal of stereochemical specificity. The sigmoidal pdG-dC binding clearly demonstrates the preference of actinomycin D for the G-C sequence. The cooperativity of the pdG-dC binding arises from the formation of Watson-Crick base pairs and strongly supports the intercalation model. The similarity of the binding curves for the pdN-dG series suggests that the orientation of guanine with respect to the chromophore is the same in all of these. It is assumed that the orientation has the 2-amino group of guanine hydrogen bonded to the carbonyl oxygen of the L-threonine residue, as described by JAIN and SOBELL (1972), in a solid-state complex of actinomycin D and deoxy- guanosine.

De SANTIS et al. (1972) have made a conformational analysis of the stereo-complex of actinomycin with DNA which should explain the high specificity of the interaction. Infrared, NMR, circular dichroism data, preliminary X-ray studies were taken as a basis to study the sterically allowed conformations of the pentapeptidelactones of acti- nomycin. The knowledge of the conformational stability of the actino- mycin structure allows the evaluation of possible distortions of the conformation, when actinomycin forms a complex with DNA.

Although having preference for the sequence G-C, actinomycin D will bind to other sequences (WELLS, 1969; WELLS and LARSON, 197o); thus

actinomycin D-DNA interaction will include a mixture of several classes of binding sites. The different dissociation constants of DNA-
actinomycin C_3 complex described by MÜLLER and CROTHERS (1968), may
correspond to different types of binding sites of DNA. It is assumed
that when actinomycin D is bound to sequences other than G-C, only
one peptide ring forms a complex with DNA, whereas when actinomycin
D is bound to a G-C sequence, both peptide rings interact with two
G's on opposite strands. If this is correct, then the most biologically active complex occurs when actinomycin D complexes to a G-C
sequence.

Actinomycin binding capacity to deoxyribonucleoproteins was studied
by RINGERTZ and BOLUND (1969). The deoxyribonucleoprotein complexes
isolated from various, mainly mammalian sources, and the corresponding DNA, were tested for binding capacity to actinomycin. While the
deoxyribonucleoprotein bound only one molecule of actinomycin per
35-14o nucleotides, the DNA preparations bound one actinomycin molecule per 14-2o nucleotides. These results indicate that deoxyribonucleoprotein binds actinomycin to a much lesser extent than DNA.
Qualitatively the actinomycin binding to deoxyribonucleoprotein closely resembles that to free DNA. The binding is sensitive to treatment
with urea, relatively insensitive to monovalent cations and is dependent on the double-helical configuration of DNA. It is therefore
concluded that the actinomycin binding of deoxyribonucleoprotein is
due to the presence of the same type of binding sites as those present in free DNA.

Binding of actinomycin to *E. coli* RNA polymerase-DNA complexes resistant to DNAase digestion was examined by BEABEALASHVILLY et al. (1973).
Actinomycin binds to this complex but does not cause its dissociation.
The apparent number of binding sites in the "complex" is approximately
equal to the value obtained for the binding of actinomycin to free
calf thymus DNA. The apparent association constant for the associate
between actinomycin and the RNA polymerase-DNA complex differs markedly from that observed for the DNA actinomycin associate. It is assumed that this reflects important structural differences between
free DNA and DNA complexed with RNA polymerase.

d) Inhibition of DNA-Dependent RNA Synthesis *in Vitro*

As a consequence of complex formation with DNA, actinomycin inhibits
the DNA-dependent RNA polymerase, whereas the DNA-dependent synthesis
of DNA with the Kornberg enzyme *in vitro* is only slightly affected
(HARBERS and MÜLLER, 1962; GOLDBERG et al., 1962; HURWITZ et al.,
1962; HARTMANN et al., 1962).

SUEOKA (1961) showed that the DNA of crabs of the genus *Cancer boreatis*
contains a poly dAT component which contains about 3% GC. This DNA
was isolated and has been used as a template for RNA polymerase (GOLD-
BERG et al., 1962, 1963; CHENG and SUEOKA, 1964) and for actinomycin
binding studies (GOLDBERG et al., 1962). Actinomycin does not bind
to synthetic dAT, but does bind to crab DNA, which contains 3% GC and
inhibitis the RNA polymerase reaction, when this crab DNA is used as
a primer.

In the RNA polymerase reaction, several sequential processes are involved: 1. association of the DNA template with the enzyme, 2. stabilization of the complex of enzyme and DNA by the nucleotide which
forms the 5'-terminus of the RNA chain, 3. initiation of the chain
by formation of the first 5'3'-linkage, 4. chain elongation, 5. lib-

eration of the synthesized polyribonucleotide chain from the template. Or, transcription may be considered as a three-step process, namely 1. initiation, 2. propagation, 3. termination. Actinomycin may have an effect at any or all of these three steps. Enzymic synthesis of RNA *in vitro* with phage T7 DNA as template was studied by RICHARDSON (1966a, b). RNA chains are initiated but the size of the RNA molecules made on the template was found to be much smaller in the presence than in the absence of actinomycin. This suggests that the bulky actinomycin molecule may prevent RNA polymerase from moving along the DNA chain. The initiation and growth of RNA chains in the DNA-dependent RNA polymerase reaction have been studied by MAITRA et al. (1967) who measured the incorporation of $\gamma^{32}P$-labeled nucleoside triphosphates and ^{14}C (or $\alpha^{32}P$) nucleoside triphosphates into RNA. The ratio of the total RNA synthesized to total initiation was used as a measure of the average chain length of the RNA produced in the polymerase reaction. A preponderance of RNA chains beginning with purine nucleoside triphosphates has been found with different types of DNA used as primers. The authors examined a number of inhibitors for their differential effects on initiation and chain growth, and found that actinomycin D barely affects the initiation of RNA chains, but markedly inhibits the synthesis of RNA. This leads to a marked decrease in the average chain length of the RNA product.

Kinetic studies by HYMAN and DAVIDSON (197o) on the *in vitro* inhibition of the transcription process by actinomycin provide evidence that actinomycin inhibits the rate of chain growth. Actinomycin inhibits propagation by affecting the rate terms for both GTP and CTP, but not for ATP and UTP. This indicates that the drug is interacting with G-C base pairs in a special way.

e) Effect on RNA Metabolism

ENGELS (1969) obtained direct evidence for the interaction of actinomycin with DNA in intact cells. The intracellular distribution of tritium-labeled actinomycin in Ehrlich ascites tumor cells was studied by HARBERS et al. (1963 a) and HARBERS et al. (1963 b) who found that it accumulated in the nuclei. KAWAMATA et al. (1965) showed by autoradiography in a human cell line that ^3H-labeled actinomycin accumulated preferentially in the nuclear region.

Microorganisms: A disproportionate production of DNA in the absence of RNA- and protein-synthesis in B. *subtilis* was first reported by SLOTNICK (1959). The preferential inhibitory effect of actinomycin on RNA synthesis in B. *subtilis* was observed by KERSTEN and KERSTEN (1962 a). ACS et al. (1963) observed that in addition to inhibiting RNA synthesis, the drug depolymerizes RNA, especially the RNA which accumulates in the presence of chloramphenicol. These results were confirmed by HANDSCHACK and LINDIGKEIT (1966). E. *coli* and B. *subtilis* respond differently to actinomycin. A permeability barrier was postulated as being responsible for the resistance of E. *coli* to actinomycin, since the drug inhibits RNA synthesis in extracts of E. *coli* (HURWITZ et al., 1962). Sensitivity to actinomycin can be produced in E. *coli* by treatment with ethylene diamine tetraacetate (LEIVE, 1965 a, b, c). Brief treatment of E. *coli* cells with EDTA renders them completely sensitive to the subsequent action of actinomycin D. Inhibition of RNA synthesis in protoplasts of E. *coli* has also been shown by MACH and TATUM (1963).

SEKIGUCHI and IIDA (1967) isolated mutants of E. *coli* which are permeable to actinomycin. In these mutants RNA synthesis can be inhibited by actinomycin. The mutants are also sensitive to lysozyme and charged molecules.

Higher plants: Besides its effect on RNA synthesis in microorganisms, actinomycin also affects the synthesis of RNA in higher plants (CHEN et al., 1968; LADO and SCHWENDIMANN, 1969).

Mammalian cells: Actinomycin inhibition of DNA-dependent synthesis of RNA in nuclear extracts of HeLa cells was observed by GOLDBERG and RABINOWITZ (1962); by SHATKIN (1962); by SALZMAN et al. (1964); by LEVY (1963). The inhibitory effect of actinomycin on RNA synthesis was also observed in Ehrlich ascites tumor cells by HABERS and MÜLLER (1962), by BASERGA et al. (1965) and by ROBERTS and NEWMAN (1966). Several investigators showed that in mammalian cells treated with moderate doses of the drug the synthesis of high-molecular-weight RNA is almost completely inhibited. A residual incorporation has been attributed to the turnover of the terminal pCpCpA sequence of tRNA through the action of a pyrophosphorylase which has now been identified (TAMAOKI and MUELLER, 1962; MERITS, 1963, 1965; FRANKLIN, 1963; HAREL et al., 1964).

In primary cultures of cell lines derived from African Green Monkey kidney (37 RC) a rapid recovery of RNA synthesis after removal of the drug was observed (BENEDETTO et al., 1972). The recovery capacity is related to the tendency of the DNA to bind less actinomycin than HeLa cell DNA and to bind a large fraction of this amount in a weaker way.

TAMAOKI and MUELLER (1962) measured the labeling of different classes of RNA with precursors in the presence of actinomycin in HeLa cells. Analysis of the labeling revealed that actinomycin D blocked the incorporation into all RNA fractions. Incorporation into the 4S class was inhibited but to a lesser extent. Since RNA polymerase is responsible for the synthesis of both sRNA and ribosomal RNA, it remains to be explained why actinomycin inhibits the synthesis of rRNA so much more than that of sRNA. The best explanation may be that of FRANKLIN (1963), who suggested that since the cistron for sRNA is much shorter than that for rRNA, sRNA will be more likely to escape the effect of actinomycin. Selective inhibition of ribosomal RNA synthesis by actinomycin has been reported by PERRY (1963), ROBERTS and NEWMAN (1966) and KAY and COOPER (1969). In the current model of ribosomal RNA synthesis in mammalian cells one molecule of 45S RNA gives rise to one molecule each of 28S and 18S RNA (DARNELL, 1968). The finding of KAY and COOPER (1969) that low doses of actinomycin preferentially inhibit the accumulation of 28S RNA is open to two interpretations: either a deficient precursor is synthesized in the presence of the drug, or the 28S RNA or its precursors are rapidly degraded. In the absence of any evidence for the formation of an altered precursor it is suggested that abnormal maturation may be the basis for the selective inhibition of ribosomal RNA synthesis by actinomycin.

Actinomycin not only inhibits the synthesis of RNA, but can also cause an artifical breakdown of RNA. This was shown in primary chick fibroblast cells in which the UTP pool has been lowered drastically by incubation with glucosamine. Under these conditions RNA catabolism can be followed by labelling experiments with [3H]-uridine (SCHOLTIS-SEK, 1972).

In addition to the unstable nucleoplasmic heterogenous RNA and the nucleolar-ribosomal-precursor RNA the nucleus of vertebrates contains a number of stable species of low molecular weight RNA components

(called small nRNA) ranging from 9o-2oo nucleotides. At least 9 components have been characterized besides tRNA, 5-S RNA and 7-S RNA.
Low concentrations of actinomycin D abolish the synthesis of 7-S rRNA
(or 28 S associated rRNA).Amounts even ten times higher are not sufficient to inhibit completely the synthesis of other small nRNAs although the synthesis of high molecular weight species of RNA are abolished (HAMELIN et al., 1973).

Several investigators have used actinomycin to calculate the half-
life of mRNA in bacteria, in plants and also in mammalian cells. These investigations are not discussed here because it is obvious that
it is extremely difficult to calculate the lifetime of mRNA in cells
which have been treated with actinomycin. As shown in the preceding
chapter, actinomycin inhibits the propagation of RNA chains, allowing
short pieces of RNA (which are probably rapidly degraded) to be made.
Serious reservations regarding the use of actinomycin D for measuring
the half-life of mRNA in mammalian cells were raised by SARMA et al.
(1969) who showed that actinomycin selectively causes degradation of
free polyribosomes of mouse liver, but that membrane-bound polyribosomes are resistant to actinomycin. Thus, under conditions where RNA
synthesis is inhibited and little or no new mRNA is being formed, it
is likely that marked disaggregation of free (but not membrane-bound)
polyribosomes may occur. Similar results were reported by BLOBEL and
POTTER (1967) and by HILL and SAUNDERS (1969). A breakdown of poly-
somes in actinomycin-treated cells was reported by STAEHELIN et al.
(1963).

Actinomycin D has been widely used to analyze the mechanism of regulation of protein biosynthesis. FIRTEL et al. (1973) studied the effect of actinomycin on the regulation of enzyme biosynthesis during
development of *Dictyostelium discoideum*. Actinomycin, when added to differentiating cells of this organism completely inhibits the synthesis
of ribosomal RNA but allows at least partially the synthesis of mRNA.
The investigators make an important statement: "Earlier experiments
using this drug, which indicated that the period of synthesis of mRNA
for several developmentally regulated enzymes preceded the period of
translation by several hours, must be re-evaluated. It is shown that
a combination of actinomycin D and daunomycin rapidly inhibits the
synthesis of mRNA; together with this the biosynthesis of the three
developmentally regulated enzymes is inhibited completely. The results
indicate, that the primary control of protein biosynthesis during
Dictyostelium development is at the level of mRNA transcription".

There is no doubt that actinomycin preferentially or even selectively
inhibits RNA synthesis within different types of cells and organisms
without inhibiting the replication of DNA. At concentrations at which
RNA synthesis is drastically inhibited, DNA synthesis can proceed
(SLOTNICK, 1959). Unter the influence of actinomycin the DNA content
in prokaryotes as well as in eukaryotes can increase about twofold
above normal levels (KERSTEN and KERSTEN, 1962 a; CLEFFMANN, 1966).

3. Particular Activities in Biological Systems

a) Synthesis of Enzymes

Actinomycin can block protein synthesis via the synthesis of RNA.
Therefore the antibiotic has been used to prevent the synthesis of
induced enzymes. The induction of β-galactosidase can be totally inhibited by actinomycin under conditions in which the over-all synthe-

sis of protein was reduced by about 60% (MOSES and SHARP, 1966). Since then actinomycin has been used to inhibit synthesis of enzymes.

However, several investigators have shown a paradoxical effect: at low concentrations, actinomycin itself can stimulate the rate of synthesis of certain enzymes, e.g. penicillinase (POLLOCK, 1963) and ribonuclease (COLEMAN and ELLIOTT, 1964). The selectivity of this effect is indicated by the observation that neither β-galactosidase nor amylase formation was stimulated under the same conditions. Similar observations have been made for enzymes in animal cells. Treatment of mice with actinomycin D resulted in an increase in intestinal alkaline phosphatase activity (MOOG, 1964, 1965). The same enzyme is stimulated by actinomycin in mammalian cells in tissue culture (NITOWSKY et al., 1964). In addition, the cortisol-inducible rat liver enzymes, alanine transaminase, tyrosine transaminase, serine dehydrase and tryptophan pyrrolase have all been shown to be induced or stimulated by actinomycin D (ROSEN et al., 1964; GARREN et al., 1964). In differentiating lens cells, actinomycin inhibits the synthesis of lens proteins in the epithelial cells and stimulates the synthesis of the same lens protein in the fiber cells (PAPACONSTANTINOU et al., 1966).

Increased rate of ornithine δ-transaminase activity in Chang's liver cells on treatment with actinomycin during the growth cycle was observed by STRECKER and ELIASSON (1966). Another stimulating effect of actinomycin was observed in rat adipose tissue by EAGLE and ROBINSON (1964) who found that low doses of actinomycin cause a marked and sustained rise in the activity of the enzyme, clearing factor lipase, in rat epididymal fat bodies, both *in vivo* and *in vitro*.

The stimulation of enzyme synthesis by actinomycin has been attributed to the inhibition of the synthesis of a repressor protein (GARREN et al., 1964). Results of SCARANO et al. (1964) also favor this view. The authors studied the control of enzyme synthesis during the early embryonic development of sea urchins: in the presence of actinomycin the embryonic development is arrested at the blastula stage, depending upon the actinomycin concentration. The embryos disaggregate and free cells remain in the culture medium. Actinomycin changes the pattern of the amount of deoxycytidylate aminohydrolase in the developing embryos; after a decrease during the first 2o hours of development, the enzyme remains constant or increases to values even higher than those of unfertilized eggs. It is suggested that actinomycin interferes with the synthesis of a specific repressor.

In this connection it is worth noting that actinomycin can stimulate the synthesis of hormone-induced enzymes in animal tissues - a phenomen known as superinduction. McCOY and EBADI (1967) reported on the paradoxical effects of hydrocortisone and actinomycin on the activity of rabbit leukocyte alkaline phosphatase: progesterone and prednisolone increased alkaline phosphatase activity while hydrocortisone caused no increase in activity; when the steroids were administered simultaneously with actinomycin D the progesterone-mediated increase was blocked. In contrast, prednisolone or hydrocortisone + actinomycin resulted in increased enzyme activity. The hormonal induction of avidine was measured *in vitro* in chick oviduct (O'MALLEY, 1967): progesterone added *in vitro* to minced chick oviduct in tissue culture medium induced the synthesis of avidine; avidine synthesis was apparent at 6 hrs and reached a maximum at 48-72 hrs. Avidine synthesis *in vitro* was prevented by actinomycin D added at zero time, but not at 6 hrs or later.

In another *in vitro* system, rat hepatoma cells, further stimulation of tyrosine-amino-transferase by actinomycin D has been shown following hormonal induction (THOMPSON et al., 197o). Certain steroid hormones induce a tenfold increase in tyrosine aminotransferase synthesis in this cell system. Actinomycin D, given after maximum induction, provokes an additional rise in enzyme activity. The same dosage of the drug almost entirely blocks the steroid-mediated induction if given to the basal cells together with or before the steroid. Evidence is presented which suggests that superinduction is a result of blocking RNA synthesis.

b) Phages and Viruses

DNA phages and viruses: E. *coli* phages T2r are not inactivated when incubated directly with actinomycin. Actinomycin, however, inhibits the reproduction of T2r on infection of E. *coli* under conditions in which the DNA synthesis of the host remains unaffected. Evidence was obtained that the phage DNA was produced, but not the coat protein. The percentage of the phage DNA formed in the presence of actinomycin remained almost constant, whereas the phage titer decreased significantly (NAKATA et al., 1961).

Under at least two different sets of experimental conditions, actinomycin can block the development of T4 phage progeny in sensitized cells of E. *coli* without measurably interfering with the synthesis of host RNA, DNA or protein (KORN et al., 1965). The mechanism by which actinomycin inhibits bacteriophage T4 maturation, however, could not be clarified. In 1967 KORN reexamined the effect of actinomycin on bacteriophage T4 maturation; replicating species of phage DNA were synthesized but their conversion to mature DNA was blocked. It was suggested that actinomycin inhibits the incorporation of DNA into head proteins or inhibits the synthesis of a constituent necessary for phage assembly. SAUER and MUNK (1966) and SAUER et al. (1966) described how treatment of herpes simplex virus-infected HeLa cells with actinomycin during the eclipse phase results in an irreversible inhibition of the multiplication of the DNA virus. Infectious virus is produced to a limited extent when the antibiotic is added during the maturation phase of the virus. It is suggested that the DNA, which becomes incorporated into these virions after treatment with actinomycin, is in a state which does not allow complex formation with the antibiotic. FISCHER and MUNK (197o) studied the effect of actinomycin on the synthesis of DNA in SV_{40} virus-infected, exponentially growing and stationary primary renal cells of *Cercopithecus*. Pretreatment with actinomycin promoted the expression of the viral genome of growing cells but not of the stationary cells. These results were explained by a preferential inhibition of cellular mRNA synthesis in exponentially-growing cells in which, in consequence, more ribosomes were present for the translation of the viral mRNA.

RNA phages or viruses: REICH et al. (1961) used actinomycin to distinguish DNA or RNA-containing viruses. They oberserved that actinomycin inhibits the synthesis of ribonucleic acids in mouse L cells and the yield of vaccinia virus-containing DNA, but does not affect the multiplication of Mengo virus-containing RNA. Mengo virus replicates normally in the presence of actinomycin at concentrations that inhibit 99.9% of the host RNA synthesis. It was therefore postulated that the biosynthesis of the RNA viruses differed from that of DNA-dependent RNA. However, some RNA viruses are sensitive to the effect of actinomycin, for example Rous sarcoma virus (TEMIN, 1963; BADER, 1964). Smaller concentrations of actinomycin inhibit the growth of some myxoviruses such as influenza (BARRY et al., 1962) and fowl plague virus (BARRY, 1964).

Actinomycin at high concentrations inhibits the replication of the phage MS2 (HAYWOOD and HARRIS, 1966; LUNT and SINSHEIMER, 1966). Reovirus differs from other small RNA-containing animal viruses in that its nucleic acid is double stranded (GOMATOS and TAMM, 1963). LOH and SOERGEL (1965) treated reovirus-infected HeLa cells with 2 µg/ml of actinomycin, which resulted in the complete inhibition of viral RNA synthesis. However, more than 8o% of these actinomycin-treated infected cells continued to produce viral antigens. The reason for this continued formation of viral protein in the absence of viral RNA synthesis is not clear. Actinomycin inhibited the formation of RNA by 9o%.

HO and WALTERS (1966) studied influenza viruses which specifically induce a ribonucleic acid nucleotidyl transferase. Actinomycin prevents the appearance of the viral enzyme in tissues infected with the influenza virus. It was suggested that the antibiotic inhibited virus multiplication by blocking the synthesis of viral RNA nucleotidyl transferase.

Since the discovery that RNA tumor viruses contain an enzyme, reverse transcriptase, which can transcribe DNA from an RNA template (TEMIN and MIZUTANI, 197o; BALTIMORE, 197o), the effect of actinomycin on the reproduction of RNA viruses can be explained. McDONNELL et al. (197o) studied in detail the effect of actinomycin on two DNA polymerases of RNA Rous sarcoma virus. Some observations suggest that there are at least two enzymatic reactions involved in viral synthesis: synthesis of DNA from a RNA template with the formation of RNA:DNA hybrid, and subsequent synthesis of double-stranded DNA. Both enzymatic activities are distributed in a homogenous fashion throughout the virion population. This confirms reports that RNA tumor viruses contain both RNA-dependent DNA polymerase and a DNA-dependent RNA synthesizing enzyme. SPIEGELMAN et al. (197o) suggested that only the DNA-dependent RNA polymerase was inhibited by actinomycin D.

c) Immune Response

Actinomycin has been used to study four immunological processes: 1. the induction of antibody synthesis, 2. immunological memory, 3. delayed hypersensitivity and 4. the rejection of transplanted tissue. An extensive review covering the literature up to 1967 is given by TANNENBERG and SCHWARTZ (1968). Actinomycin appears to have a significant inhibitory effect on the induction of the primary antibody response and very little effect on the acquisition or expression of immunological memory. In clinical immunology actinomycin can be used only temporarily in the acute therapy of the rejection crisis that follows renal homotransplantation. There is little evidence as yet to show that these empirical effects are related to specific interference with the immune response. Because of the severe systemic toxic effect produced by actinomycin, its use in other immunological disorders is limited.

B. Anthracyclines

1. Origin, Biological and Chemical Properties

According to BROCKMANN et al. (1963) antibiotics that contain a tetrahydrotetracenequinone chromophore linked to a sugar side chain are named anthracyclines. *Daunomycin* is a metabolite of *Streptomyces peucetius* (GREIN et al., 1963). In 1964 DUBOST et al. isolated *rubidomycin* from

Streptomyces coeruleorubidus. This was later found to be identical with
daunomycin. The drug, widely used for the treatment of cancer, has
been given the name *daunorubicin. Cinerubines*, namely A and B, are metabolic products of actinomycetes (ETTLINGER et al., 1959). *Ruticulomycins*
are produced by *Streptomyces rubireticuli* (MITSCHER et al., 1964). *Nogala-*
mycin was purified by BHUYAN and DIETZ (1965) from *Streptomyces nogalater.*
Adriamycin was isolated in 1967 by the FarmItalia Research Laboratories
from cultures of a mutant *Streptomyces peucetius* var. *caesius.* The anthracycline antibiotics are active against gram-positive bacteria and
fungi. Gram-negative bacteria are generally more resistant to the anthracyclines. The anthracycline antibiotics possess antiviral activity and strongly inhibit the growth of a variety of experimental tumors.

The chemical structures of daunomycin and related antibiotics are
shown in Figures 31-34. The structure of the chromophore of daunomycin - daunomycinone - and the structure and stereochemistry of the
attached sugar - daunosamine - were elucidated by ARCAMONE et al.
(1964 a, b). The amino sugar is attached to ring D in position 7 (IWA-
MOTO et al., 1968). The structure of adriamycin differs from that of
daunomycin only in the substitution of hydrogen atom with the hydroxyl group on the acetyl residue (ARCAMONE et al., 1969).
An X-ray analysis of the N-Br-acetyl derivative of daunomycin carried
out by ANGIULI et al. (1971) without any assumption about the structure of the daunomycin molecule, gave excellent correlation with previous chemical studies and established the stereochemistry though not
the absolute configuration. The cyclohexene ring is in the "half chair"
conformation. The angle between the least squares planes passing through
the sugar ring and the chromophore is 1oo°, a value similar to those
found for other molecules with two more or less planar systems.

The structure of cinerubine A was elucidated by KELLER-SCHIERLEIN and
RICHLE (197o). Cinerubine A is a glycosidic anthracycline antibiotic
whose aglycone is ε-pyrromycinone. The sugar side chain consists of three
components: L-rhodosamine, 2-deoxy-L-fucose and an unusual ketosugar,
L-cinerulose A (2,3,6-trideoxy-L-aldohex-4-ulose). The three sugars
are arranged in a trisaccharide chain attached to the oxygen atom in
position 7 of the aglycone. The only anthracycline antibiotic with
more than one sugar side chain, whose structure determination is practically complete, is β-rhodomycin, whose trisaccharide side chain corresponds largely with that of cinerubine A (KELLER-SCHIERLEIN and

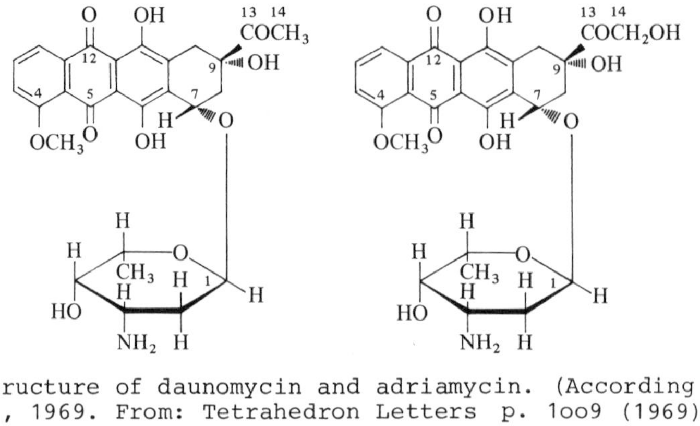

Fig. 31. Structure of daunomycin and adriamycin. (According to ARCA-
MONE et al., 1969. From: Tetrahedron Letters p. 1oo9 (1969))

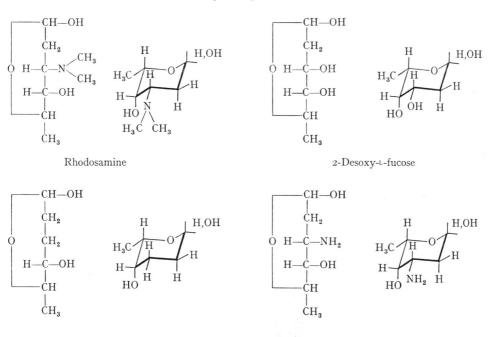

	R_1	R_2	R_3	R_4	R_5	R_6	R_7
ε-Pyrromycinone	OH	OH	H	OH	COOCH₃	OH	C₂H₅
Aclavinone	H	OH	H	OH	COOCH₃	OH	C₂H₅
β-Rhodomycinone	H (or OH)	OH (or H)	OH	OH	OH	OH	C₂H₅
γ-Rhodomycinone	H (or OH)	OH (or H)	OH	OH	OH	H	C₂H₅
β-Isorhodomycinone							
5-Desoxy-pyrromycinone	OH	OH	H	H	COOCH₃	OH	C₂H₅
Daunomycinone	H (or OCH₃)	OCH₃ (or H)	OH	OH	H	OH	COCH₃

Sugars components

Rhodosamine

2-Desoxy-ʟ-fucose

Rhodinose

Daunosamine

Fig. 32. Structures of anthracyclines. (According to DI MARCO, 1967. From: Antibiotics I. Eds.: GOTTLIEB, D., and SHAW, P.D. Berlin-Heidelberg-New York: Springer 1967, p. 192)

RICHLE, 197o). The unusual sugars "uloses" with a keto group in position 4 were unknown as natural products until recently. Now these sugar components have been found as building blocks of certain macrolide antibiotics which exhibit quite different chemical behavior and mode of action than the hitherto known macrolides (SUZUKI, 197o). Ruticulomycins contain as chromophore 5-deoxy-pyrromycinone, with as

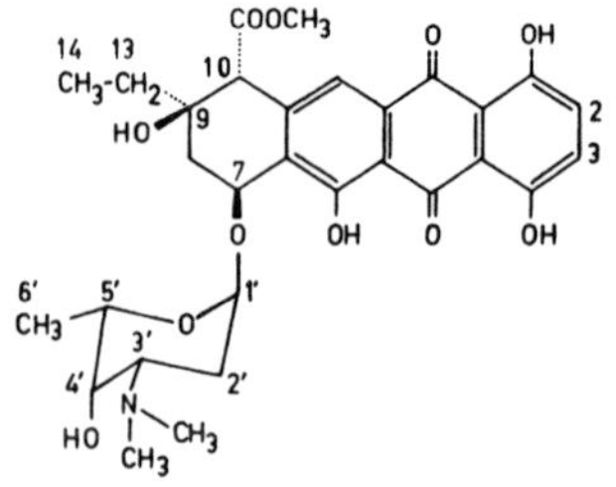

Pyrromycin

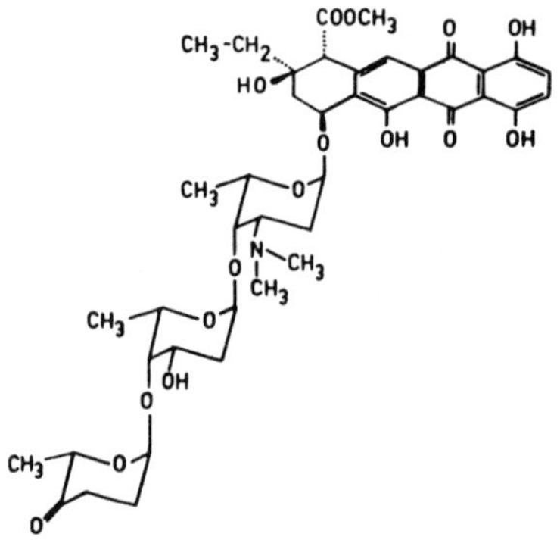

Cinerubine A

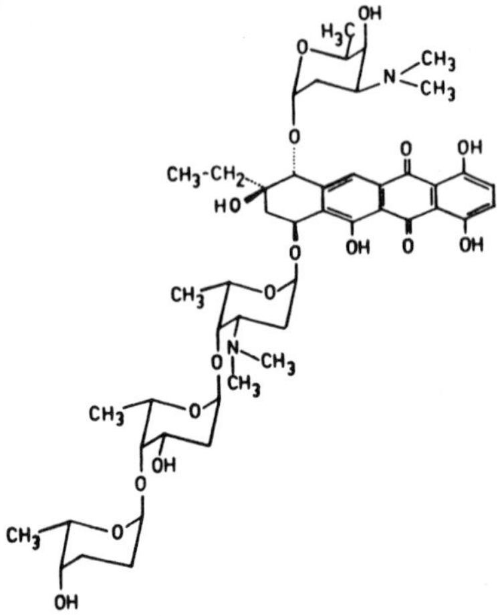

ß-rhodomycin

Fig. 33. Structures of pyrromycin, cinerubine A and ß-rhodomycin.
(According to KELLER-SCHIERLEIN and RICHLE, 197o. From: Antimicrobial
Agents and Chemotherapy 197o, p. 7o, 73)

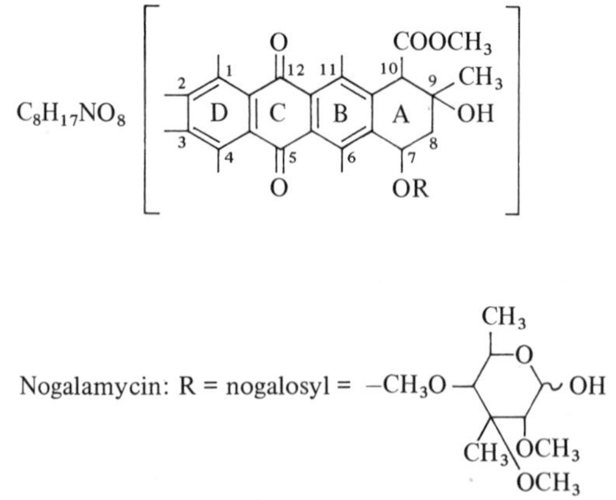

$C_8H_{17}NO_8$

Nogalamycin: R = nogalosyl = $-CH_3O$

Fig. 34. Structure of nogalamycin and analogs. (According to WILEY
et al., 1968. From: Tetrahedron Letters p. 667/668 (1968))

yet unknown sugar residues. The structure of nogalamycin and its analogs was described by WILEY et al. (1968).
Carminomycin another antitumor antibiotic of the anthracycline group
was isolated by GAUSE et al. (1973) and BRAZHNIKOVA et al. (1973 a)
from *Actinomadura carminate sp. nov.* Carminomycin I is a glycoside which
resulted upon mild hydrolysis of carminomycins 2 and 3. The amino-
sugar is daunosamine. The aglycone, carminomycinone differs from all
hitherto known anthracyclines by the presence of a hydroxy group at
C_4 and from differences in the specific optical rotation (BRAZHNIKOVA
et al., 1973b).

2. On the Molecular Mechanism of Action

a) Interaction with DNA *in Vitro*

The formation of complexes of daunomycin and of cinerubine with DNA
was shown by KERSTEN and KERSTEN (1965). Both antibiotics sediment together with DNA in the ultracentrifuge, causing a decrease in the sedimentation coefficient of DNA. The DNA-anthracycline complexes exhibit higher melting temperatures than DNA itself. The interaction
with DNA of daunomycin and cinerubine is accompanied by changes in
the spectra of the dyes. When excited at 49o nm cinerubines and rhodo-
mycins have a light emission at 57o nm. On addition of DNA a change
of the fluorescence spectra occurs (CALENDI et al., 1965).

Further evidence for the binding of daunomycin to DNA was given by
the disappearance of the waves of polarographic reduction which are
characteristic of free anthraquinones (CALENDI et al., 1965).

A detailed analysis of the physicochemical properties of complexes
between deoxyribonucleic acid and anthrycyclines, such as cinerubine,
daunomycin and nogalamycin, was carried out by KERSTEN et al. (1966).
The following physicochemical parameters of several types of native
DNAs, which vary in their GC content between 3o% and 7o%, were altered
in the presence of the anthrycyclines: thermal transition (Fig. 35),
viscosity (Fig. 36), sedimentation and buoyant density in CsCl or
Cs_2SO_4 gradients (Fig. 37). From these results one may conclude that
the interaction of anthracyclines with DNA is accompanied by profound
conformational changes of the DNA and has several features in common
with those reported for acridine dyes and similar planar compounds,
e.g. ethidium bromide (for further details see the following section).
LERMAN (1961, 1963) postulated that the dyes become intercalated between adjoining nucleotide pairs of double helical DNA. Anthracyclines
caused an increase in viscosity and a decrease in sedimentation of
native DNA, which is consistant with the "intercalation hypothesis".
The high ionic strength of the medium has little effect on the binding
of the anthracyclines to DNA, whereas at increasing salt concentrations
the acridines are fully displaced. It is suggested that the amino sugar residues are responsible for this stabilization of the antibiotic-
DNA complex, because CALENDI et al. (1965) showed that N-acetylation
of the sugar residue in daunomycin results in a profound decrease in
its affinity for DNA.

The anthracycline-DNA complexes retain their stability even at ele-
vated temperatures and at very high ionic strength. The stability of
the complex under the latter conditions contributes to the progressive decrease in the buoyant density gradients. The most pronounced
density shift was obtained with nogalamycin, which contains two sugar
residues with progressively fewer effects observed for cinerubine

(three sugar residues) and daunomycin (one sugar residue). Furthermore, the experiments of KERSTEN et al. (1966) indicate that the DNA-
anthracycline interaction increases with rising G+C content of the
DNA, although even poly dAT shows some anthracycline binding, as can
be inferred from the density shift. By the same criterion denatured
DNAs seem to bind anthracyclines to only a slightly lesser degree than
native DNA. However, viscosity augmentation and decrease in sedimentation constant were observed only for anthracycline complexes with
native but not with denatured DNA.

ZUNINO (1971) studied the intrinsic viscosity of sonicated native DNA
in the presence of daunomycin at different molar ratios of drug: nucleotide. He calculated the number of binding sites which were found
to correspond closely to the apparent number (o.12) of binding sites
per nucleotide obtained in a spectrophotometric titration of daunomycin with DNA, and to the molar ratio drug: nucleotide of about o.12-
o.16 recovered by an urea gradient in a DNA-cellulose column.

COHEN and EISENBERG (1969) concluded that changes in hydrodynamic properties of sonicated DNA fragments arising from the intercalation of
dye molecules are probably almost exclusively due to an increase in
contour length with respect to viscosity. The viscosity enhancement
for the daunomycin-DNA complex is significantly larger than that suggested by ULLMAN's calculation (1968) for intercalating dyes. For this
reason the high daunomycin-induced viscosity increase could not be

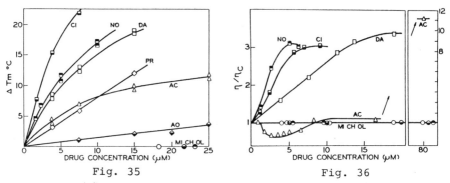

Fig. 35   Fig. 36

Fig. 35. Effect of antibiotics and acridine dyes on the shift in the
melting temperature (ΔT_m) of *C. johnsonii* DNA (4o µg of DNA/ml of PE
solvent). The following symbols, abbreviations of drug names, and
molecular weights were used for calculating the molar concentrations
or ratios: actinomycin D (C_1), *AC* (△), 12oo; chromomycin A$_3$, *CH* (●),
1o52; mithramycin, *MI* (◐), 11oo (RAO et al., 1962, and RAO, K.V.,
personal communication); olivomycin, *OL* (○), 96o; cinerubin, *CI* (◨),
875 (ETTLINGER et al., 1959); daunomycin, *DA* (□), 56o; nogalamycin,
NO (◪), 793; acridine orange, *AO* (◆), 3o1; and proflavine sulfate
(.H$_2$O), *PR* (◇) 324; 1 µM = 1 µg/ml for each 1ooo molecular weight
values

Fig. 36. Effect of several antibiotics at various concentrations on
the relative increase in the reduced specific viscosity (η/η_c) of
C. johnsonii DNA (2o µg/ml of PE solvent), measured at 25° and at
o.o1-o.o5 sec^{-1} shear rate. Reduced specific viscosity of control
DNA solution η_c = 16o dl/g. Viscosity of DNA + antibiotic designated
as η

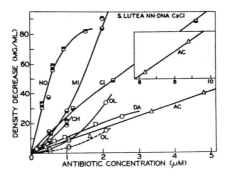

Fig. 37. Effect of various antibiotics (mμM/ml of CsCl solution) on
the decrease in buoyant density (control density minus density in
presence of antibiotics) of native (NN) *S. lutea* DNA in CsCl gradient.
DNA (2 μg) was well mixed with the indicated amounts (mμM) of antibiotics in o.2 ml of PE buffer and 2-1o min later supplemented with
o.8 ml of saturated CsCl solution. The density of the mixture was
adjusted to approximately 1.7 g/ml. Centrifugation was carried out
for 22-24 hr at 44.77o rpm, 25°. The density determinations are based either on the micropicnometric measurements with an accuracy of
o.1 μg/ml (anthracyclines) or on the band position with respect to
the poly dAT reference band, the density of the latter found not to
be affected by the other antibiotics (MI, CH, OL, AC) tested under
the present experimental conditions. Since addition of olivomycin
causes splitting of the DNA band in addition to a density decrease,
the solid line indicates the shift in the more affected component
and the broken line in the less affected one. The insert at the upper
right corner provides density shift data for the higher actinomycin
concentration. For curve designation see legend to Fig. 35

Figs. 35-37. Physicochemical properties of complexes between deoxyribonucleic acid and antibiotics which affect ribonucleic acid synthesis (actinomycin, daunomycin, cinerubin, nogalamycin, chromomycin,
mithramycin and olivomycin). (According to KERSTEN et al., 1966. From:
Biochemistry 5, 238-24o (1966))

completely explained by an increase in the contour length of the DNA
chain. Other effects, such as a decrease in the flexibility of the
DNA by daunomycin, are possible.

The anthracyclines propably differ from actinomycins with respect to
the binding sites in DNA. RUSCONI (1966) observed that mercury ions
act differently on the two antibiotic-DNA complexes. While actinomycin
is released from DNA upon the addition of Hg^{2+}, daunomycin is not separated from DNA even at high concentrations of Hg^{2+}. That the binding sites in DNA for actinomycin and daunomycin are at least partially
distinct, was also concluded from the fact that DNA treated in excess
with actinomycin is still able to interact with daunomycin. ZUNINO
(1971) studied the mode of interaction of daunomycin with DNA by using
DNA-cellulose column chromatography according to INAGAKI and KAGEYAMA
(197o). During the elution procedure daunomycin was eluted at about
o.2-o.4 M in a NaCl gradient and at about 2.5-3 M in a urea gradient
(Fig. 38). The chromatographic behavior could not be attributed to
a mixture of two different molecular species of daunomycin. The results suggest that daunomycin exhibits a mode of interaction with DNA

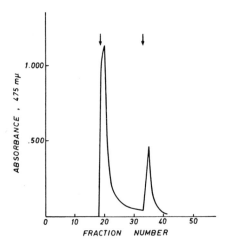

Fig. 38. DNA-cellulose chromatography of daunomycin. o.2 ml of o.5% daunomycin solution (in o.o1 M Tris, o.oo1 M EDTA, pH 7.o) were loaded on a o.6 x 4 cm column. The daunomycin loading took place at fraction ,O. After loading, the column was washed with the buffer and the bound material was eluted by 2 M NaCl and succesively by 7 M urea in 2 M NaCl-o.o1 Tris-o.oo1 M EDTA, pH 7.o. The stepwise elutions are indicated by the vertical arrows. Fractions of 3.7 ml were collected. Recovery was 1oo%. Flow rate, 37 ml/hr. (According to ZUNINO, 1971. From: FEBS Letters 18, 25o (1971))

involving binding (probably electrostatic) in which the DNA phosphate groups and the daunomycin amino-group participate, and a stronger binding mode requiring urea for the dissociation.

Circular dichroism (CD) spectra of complexes of DNA with daunomycin and with nogalamycin were measured (FEY, 1973; DALGLEISH et al., 1973; see Fig. 39). From a matrix rank analysis it is suggested that the CD spectra are composed of 4 components, indicating that besides free daunomycin and the altered DNA, two DNA-daunomycin complexes are present. Also the anthracycline antibiotic nogalamycin appeared to require four components to fit the complex CD spectra. The observation that the CD spectra of the DNA-antibiotic complexes differ qualitatively and qantitatively suggests that nogalamycin and daunomycin bind to DNA in a different way, probably as a result of their differences in the sugar residues (nogalamycin contains two and daunomycin one sugar residue).

The complex formation between daunorubicine and native or heat-denatured DNA was also studied by BARTHELEMY-CLAVEY et al. (1973) using dialysis equilibrium, spectrophotometry and thermal denaturation. It was found that two types of complexes are formed with both native and denatured DNA.

b) Interaction of Planar Dyes with DNA and tRNA *in Vitro*

DNA: The physicochemical properties of the anthracycline-DNA complexes resemble in many respects the physicochemical properties of complexes between DNA and planar dyes such as aminoacridines, ethidium bromide,

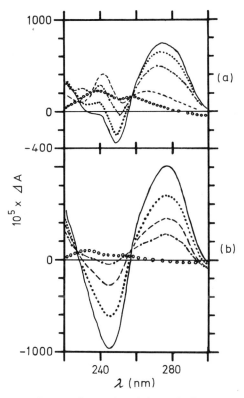

Fig. 39. CD spectra of nogalamycin (a) and daunomycin (b) with increasing concentrations of DNA from calf thymus. Nogalamycin 8 x
10^{-5}M, free drug ——; drug to DNA-P (= D/P) molar ratio o.o38 ....;
o.o48 -.-.-.-; o.o76 - - -; o.114 ooooo. Daunomycin 6.8 x 10^{-5}M,
free drug ——; D/P = o.o193 .....; o.o23 ----; o.o37 -.-.-.-;
o.o965 ooooo. (According to FEY, 1973)

some antimalarials, e.g. chloroquine or quinacrine (WARING, 197o, 1971;
for further review see Progress in Molecular and Subcellular Biology,
Vol. 2, 1971, E. HAHN, Editor). Some studies on the interaction of
dyes with DNA are presented in this section because the knowledge of
the nature of the complexes between dyes and DNA may contribute to
our understanding of the anthracycline antibiotic-DNA interaction.

Several models have been suggested for the dye-DNA complexes:

1. *The stacking model* (BRADLEY and WOLF, 1959). This model (Fig. 4o)
proposes the binding of numerous ring systems of positively charged
dyes to the periphery of the double helix by electrostatic attraction
to the negatively charged phosphate groups of DNA. This type of binding is characterized by a stoichiometry which exceeds one ligand molecule per 4 or 5 component nucleotides.of DNA and by comparatively low
apparent binding constants. These binding processes are antagonized
by monovalent, and more strongly antagonized by divalent inorganic
cations. The neutralization of DNA phosphates by positively-charged
counterions stabilizes DNA to strand separation and in consequence

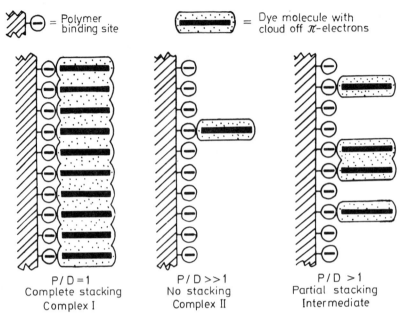

Fig. 4o. Diagram of dye stacking on polymer. (According to BRADLEY and WOLF, 1959)

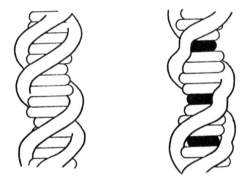

Fig. 41. Intercalation into DNA. The distortion of DNA structure caused by an intercalating drug is shown. Left: normal DNA. Right: DNA containing intercalated drug molecules. (According to WARING, 197o)

increases the thermal denaturation temperature of DNA. See also BRADLEY and LIFSON, 1968.

2. *The intercalation model* according to LERMAN (1964) proposes the insertion or intercalation of flat aromatic ring systems between the levels of base pairs into the DNA double helix (Fig. 41). This type of binding is characterized by a stoichiometry not exceeding one ligand molecule per two base pairs. As a prerequisite to fit this mo-

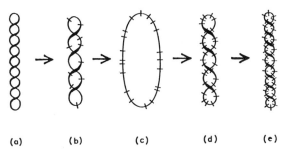

(a)   (b)   (c)   (d)   (e)

Fig. 42. Intercalation into supercoiled DNA. A representation of the progressive removal and reversal supercoiling turns associated with the binding of increasing amounts of an intercalating drug. (According to WARING, 1970. From: Naunyn-Schmiedebergs Arch. Pharmak. exp. Path. 259, Fig. 41, p. 92; Fig. 42, p. 93 (1970))

del the DNA molecule as a whole must become extended to provide the spaces for the drug to be inserted. Local uncoiling must occur before these spaces can appear. Thus if the drug is intercalated, the DNA molecule must become extended and partially uncoiled.

A local uncoiling of DNA was deduced from results of experiments carried out by CRAWFORD and WARING (1967), BAUER and VINOGRAD (1968, 1970, 1971), and WARING (1970, 1971). Supercoiled circular DNA from polyoma virus has been isolated and shown to interact with several planar dyes in increasing concentrations. The sedimentation coefficient of polyoma DNA first falls and then rises again when increasing amounts of the drug become bound to the DNA. The sedimentation coefficient of this DNA was taken as a sensitive indication of its supercoiled state (supercoils make the whole molecule more compact, so that it sediments faster in a high gravitational field). The measurements of CRAWFORD and WARING (1967) indicate that the supercoils were first removed and subsequently reintroduced as the level of the drug binding further increased. The events are represented schematically in Figur 42. The starting material a) is supercoiled DNA double helix with right-handed supercoils. As drug molecules (see bars perpendicular to the axis of the double helix) bind to the DNA, the strain in the molecule is lessened and the number of supercoils diminishes b). At a certain ratio of drug to DNA the accumulated untwisting caused by the bound drug molecules just balances the initial shortage of turns in the circular DNA double helix c). As more drug molecules bind, the circle becomes strained in the opposite direction causing left-handed supercoils to appear d) and e). These results precisely fulfill the prediction of what should happen if a drug which causes local uncoiling of the DNA helix binds to a supercoiled circular DNA molecule whose supercoils were due to a deficiency of winding turns in its double helix.

"Does intercalation wind or unwind the DNA helix?" is the subtitle of a paper by PAOLETTI and LE PECQ (1971). Theoretical data for the fluorescence depolarization of ethidium bromide bound to DNA have been obtained through computer simulation of energy transfer and have been compared with experimental determinations. This procedure permitted an evaluation of the winding angle of the DNA helix caused by intercalation. This approach to the solution of the preceding question requires some assumptions - mainly that the fluorescence depolarization due to internal motion in DNA is independent of the fluorescence depolarization due to energy transfer. The presented results seem to be in accordance with this assumption. Internal motion in DNA is relatively slow, and it is believed that bases separated by short distances move virtually synchronously.

The next assumption is that ethidium bromide does not change its fluorescence depolarization by interfering with the internal motion of DNA. The computer simulation of energy transfer between ethidium bromide molecules bound to DNA permits, then, the calculation of the theoretical depolarization of the fluorescence emitted by ethidium. Since the energy transfer is very dependent on orientation, the angle between two ethidium molecules, and therefore the local change in the winding angle of the DNA helix caused by intercalation, can be determined. The authors state that the DNA helix is wound by $13^O \pm 4^O$ and not unwound after the intercalation of one ethidium bromide molecule. In accordance with the results, appropriate models were constructed. The authors state that it is as easy to build a model identical to the one presented by FULLER and WARING (1964), which leads to a 12° unwinding, as to build a model with a 12° winding. The model can be built by a rotation of the C-5 of the sugar and of the phosphorus in the same direction as the helix (right-handed). One feature of this model is that the negative charge of the phosphate can interact with the positive charge of ethidium bromide carried on the nitrogen in position 8.

3. *The modified intercalation model* of BLAKE and PEACOCKE (1968). The modification proposed by these authors is that when acridine cations are bound by the strong binding process, which is at least in part electrostatic, the acridine lies between successive nucleotide bases on the same polynucleotide chain in a plane approximately parallel to the base planes but at an angle (looking down the polynucleotide chain) such that the positive ring nitrogen is close to the polynucleotide phosphate group (Fig. 43). This model may explain the interaction of dyes and anthracyclines not only with native double-helical DNA, but also with denatured disordered DNA. Thus this modified intercalation model does not make the presence of intact double helices a condition for the process of strong binding. A second binding process may occur in complex formation. It is weak and involves interaction between bound aminoacridine molecules; it is also electrostatic and probably a more external binding process. Denaturation of DNA enhances the tendency to bind to DNA by this second mode of interaction.

CHAN and BALL (1971) studied the binding of dimethylbenz(c)acridines to DNA and the physicochemical properties of these DNA complexes. The data obtained tend to support the modified intercalation model for the following reasons: firstly the dimethylbenz(c)acridine molecules are in the cationic form when bound to DNA. Therefore it appears that there is electrostatic interaction between the acridinium cation and the DNA phosphates. This could not happen if the dimethylbenz(c)acridine was located entirely within the interior of the helix. Secondly, the strong dependence of binding on ionic strength, the concentration of divalent cations and pH are also indicative of the existence of electrostatic interactions between the dye of molecules and the DNA bases. These substances bind to denatured DNA almost as strongly as to native DNA. Therefore the second modified intercalation model is preferred to explain the binding of these substances to DNA.

SAKODA et al. (1972) measured the kinetics on the acridine orange-DNA interaction. The results suggest a branched mechanism of binding involving intercalation and outside dimerization. AKTIPIS and KINDELIS

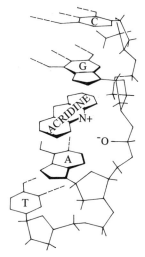

Fig. 43. Modified intercalation model for the complex of proflavine with polynucleotide chains. (According to BLAKE and PEACOCKE, 1968. From: Biopolymers 6, 1225 (1968))

(1973) investigated the ethidium bromide-DNA complex. The circular dichroism of the complex was studied at different temperatures and different ionic strength. The results indicate that at high salt concentrations the complex behaves as predicted from classical models of stacked chromophores. Direct interactions between ethidium bromide molecules intercalated into neighboring binding sites of DNA are assumed to occur.

These few examples show the difficulty of a physical and theoretical approach to problems such as the interactions of small molecules with macromolecules. Nevertheless these studies are most important and will help us in the future to obtain more precise insight into molecular associations in biology. (For review see PULLMAN, 1968).

It is generally accepted that several cationic planar dyes interact with DNA through two models of binding: intercalation between DNA bases and binding to the outside of DNA, depending on the ratio of DNA phosphate to dye (P/D ratio). To discriminate the two modes of binding, kinetic studies by the fluorescence stopped-flow method and concentration dependence of relaxation times in temperature jump studies have more recently been applied (AKASAKA et al., 197o; LI and CROTHERS, 1969).

PIGRAM (1972) published X-ray diffraction pattern of fibers of DNA-daunomycin complexes at 92% relative humidity and various phosphate-to-drug ratios. Oriented diffraction patterns showing layer line structure and resembling semi-crystalline B-DNA were obtained from a series of fibers for which the drug phosphate ratio ranged from o.o3 to o.2, as well as from control fibers with no drug. The changes which were observed in the diffraction patterns of the complexes were consistent with one out of three hypothetical models which involve intercalation; in this model the amino group of daunomycin is in the large groove of the DNA and the hydrophobic faces of the base pairs

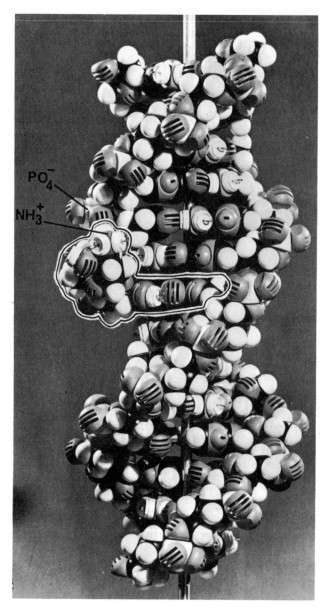

Fig. 44. Model of the DNA-daunomycin complex. (According to PIGRAM et al., 1972. From: Nature New Biology 235, 19 (1972))

and the drug overlap extensively. The amino sugar is at the side of the groove, close to a sugar phosphate chain, thus enabling the ionized group to interact strongly with the second DNA phosphate away from the intercalation site (Fig. 44). A possible additional interaction would be a hydrogen bond between the first phosphate and the hydroxyl group attached to the saturated ring of the daunomycin chromophore.

tRNA: Anthracyclines, as well as the planar dyes, interact with tRNA as with double-stranded, native and partially denatured, DNA, (BITTMANN, 1969; CHURCHICH, 1963; CANTOR et al., 1971; DOURLENT and HÉLÈNE, 1971; TRITTON and MOHR, 1971; GROSJEAN et al., 1968; FINKELSTEIN and WEINSTEIN, 1967; LURQUIN and BUCHET-MAHIEU, 1971).

Temperature-jump relaxation methods have been employed to study the kinetics of interaction between unfractionated tRNA from yeast and ethidium bromide (TRITTON and MOHR, 1973). At 25° three relaxation processes can be detected which on the basis of their concentration dependence have been ascribed to a bimolecular outside binding followed by an intercalation step and finally a fast isomerization of the intercalated dye-tRNA complex. The isomerization vanished at $1o^O$ and 15° probably because of the greater stability of tRNA structure. Aminoacylation alters the kinetics of ethidium binding especially in the bimolecular step, indicating that uncharged tRNA and aminoacylated tRNA have different conformations. Rate- and equilibrium constants, activation energies and thermodynamic parameters have been calculated for the elementary steps. At high ionic strength dye-tRNA complexes appear to dimerize.

URBANKE et al. (1973) have used ethidium bromide as a probe of tRNA conformation. tRNA melting profiles can be accounted for by several melting effects. However the assignment of these effects to definite structural elements has not yet been accomplished. The influence of ethidium bromide on the thermal denaturation profiles of tRNA^{phe} (yeast) and tRNA^{val} *(E. coli)* was investigated. The results indicate that the dye destabilizes one element of the tRNA structure and stabilizes other elements. Since ethidium bromide stabilizes double stranded regions it can be used to distinguish tertiary from secondary structure elements in tRNA.

Aminoacridines and related compounds interact with rat liver ribosomes and affect their conformation (HULTIN, 197o). BORISOVA and MINYAT (1969) showed complex formation between deoxynucleoprotein and acridines.

c) Effect on DNA-Dependent Processes *in Vitro*

The influence of the anthracyclines daunomycin, cinerubine A and B and rhodomycin on the DNA-dependent synthesis of DNA and RNA *in vitro* was measured by using the DNA polymerase and the RNA polymerase from *E. coli* (KOSCHEL et al., 1966). The anthracyclines were found to inhibit both the DNA-dependent synthesis of DNA and the DNA-dependent synthesis of RNA. In these assays DNA-dependent synthesis of DNA with the Kornberg enzyme is inhibited by daunomycin, nogalamycin, and cinerubine at a molar concentration ratio of o.25 antibiotic molecules per 1o DNA nucleotides, whereas DNA-dependent synthesis of RNA is inhibited by the anthracyclines at a molar concentration ratio of o.o3 antibiotic molecules per 1o DNA nucleotides (for review see KERSTEN and KERSTEN, 1967).

When *E. coli* RNA polymerase was used with poly dAT as a primer, nogalamycin caused 99% inhibition of RNA synthesis, as compared to only 3% inhibition when poly dG:dC was used as a primer (BHUYAN and SMITH, 1965; BHUYAN, 1967).

The inhibition of RNA synthesis by different nogalamycin analogues was compared with that using poly dAT as a primer. The amounts of

drug needed for 5o% inhibition of RNA synthesis were as follows:
(nmoles/ml): nogalamycin, o.85; O-methyl-nogalarole and nogalarene,
14.5; 7-deoxy-nogalarole, 39; and nogalarole, 51 (BHUYAN and REUSSER,
197o).

Other DNA-dependent processes, such as the methylation of methyl-de-
ficient DNA with crude methylating enzymes from *E. coli* and the degradation of DNA by DNAase I (endo-DNAase) and DNAase II (exo-DNAase),
can be inhibited by the antibiotics in cell-free systems (KERSTEN and
KERSTEN, 1968). The RNA polymerase activity was also inhibited by
ethidium bromide and proflavine. In all cases the inhibitory effect
could be overcome by the addition of DNA (WARING, 1965), showing that
the target of the inhibition is the DNA template and not the polymerase.

Using spectrophotometric methods DOSKOČIL and FRIČ (1973) compared
the complex-forming ability of daunomycin with double-stranded RNA
(replicative form of f2 phage RNA) and ribosomal RNA and found that
in contrast with DNA the secondary structure of RNA was irrelevant
for the stability of the complex and no evidence of intercalation
was obtained.

d) Effect on Nucleic Acid Metabolism

ROTH and MANJON (1969) presented evidence of a specific assiociation
between acriflavine and DNA in intact cells. It is well known that
acridines bind to intracellular nucleic acids - especially to DNA
(for review see RINGERTZ et al., 197o; LÖBER, 1971).

1. Dependence on Uptake and Cell Cycle

Although direct association of anthracyclines to DNA within intact
cells has not yet been demonstrated, it is generally accepted that
the inhibitory effect of anthracyclines on nucleic acid, especially
on RNA synthesis, is the consequence of the binding of these substances to DNA, as proposed by KERSTEN and KERSTEN (1966), DI MARCO (1967),
and DI MARCO et al. (1965). RUSCONI and DI MARCO (1969) studied the
inhibition of nucleic acid synthesis by daunomycin and its relationship to the uptake of the drug in HeLa cells. Increasing concentrations of the drug caused a proporticnally higher inhibition of the
incorporation of nucleic acid precursors. The relative inhibition of
^{3}H-uridine and ^{3}H-thymidine incorporation appeared to be dependent on
the concentration of the drug in the medium and on the duration of
contact with the cell. The uptake of the antibiotic by HeLa cells was
studied by means of tritiated daunomycin. The uptake was rapid and
virtually complete within 2o minutes. The amount fixed per cell seemed to be dependent only on the drug concentration in the medium. A
quantitative relationship between the inhibition of uridine incorporation and the uptake of the antibiotic was demonstrated. BREMERSKOV
and LINNEMANN (1969) studied the effect of daunomycin on nucleic acid
synthesis in synchronized cultures of mouse fibroblasts and showed
that during the first 12 hours of treatment, DNA and RNA synthesis
are inhibited to the same extent. Thereafter RNA synthesis is resumed,
while DNA synthesis is selectively inhibited.

SILVESTRINI et al. (197o) studied the influence of daunomycin on nucleic acid synthesis in different stages of the cell cycle in mammalian cells. Explants from leg muscle of newborn rats were synchronized
by two 24-hour treatments with excess thymidine at an interval of 16
hours. The metabolic activities of DNA and RNA were determined by

autoradiographic methods, which measured the incorporation of ^{3}H-thymidine, ^{3}H-cytidine and ^{3}H-uridine into nuclear structures. In control cultures, DNA synthesis shows two distinct waves located in the early and late S-phase. It is most sensitive to the inhibitory effect of daunomycin during the latter period. RNA synthesis takes place in the cells during the entire cell cycle from the end of mitosis to the late prophase, but two well-defined peaks can be recognized: the first in the G_2 phase, just 1 or 2 hours before the mitotic peak and the second in the middle stages of G_1. These RNA syntheses - involving both nucleolar and extranucleolar structures - are strongly inhibited by daunomycin. Thus, daunomycin produces different effects according to the physiological activity of the cells at the time of treatment. KIM et al. (1968) observed strong toxicity of daunomycin during the phase of DNA synthesis in HeLa cells in 15-18 hr after mitosis.

2. Specificity of the Effect

Selective inhibition of ribosomal RNA synthesis in HeLa cells by nogalamycin A was reported by ELLEM and RHODE (197o). A concentration of 1.39 x 1o^{-7} M was found to produce close to 5o% inhibition of cytidine incorporation into the nucleic acids of HeLa cells; 3o times this concentration had no significant immediate effect on protein synthesis. At a concentration of 1.39 x 1o^{-7} M, nogalamycin inhibited the synthesis of all types of nucleic acids in increasing order of magnitude: DNA>DNA-like RNA>tRNA>5-S RNA>rRNA. The selectivity for rRNA was pronounced. The effect of nogalamycin on the synthesis of nucleic acids in intact animals was studied in detail by GRAY et al. (1966). The results of these studies indicate that nogalamycin inhibits the synthesis of rat liver RNA *in vivo* including messenger RNA. This conclusion rests upon the assumption that precursor incorporation is an effective measure of synthesis. It must be pointed out, however, that changes in pool sizes or in degradation rates could have been effected by nogalamycin administration. Although nogalamycin inhibited RNA synthesis as measured by orotate incorporation, it did not inhibit total protein synthesis either *in vivo* or in cultured mammalian cells (BHUYAN and DIETZ, 1965). The inability to inhibit totally the synthesis of protein is explained by a relative stability of messenger RNA in mammalian cells (REVEL and HIATT, 1964). Hydrocortisone-induced synthesis of tryptophan pyrrolase in rats is inhibited by nogalamycin, indicating that nogalamycin blocks the synthesis of new messenger RNA. In microorganisms in which the messenger RNA has a relatively short half-life, nogalamycin markedly inhibits protein synthesis.

CROOK et al. (1972) also observed a selective inhibitory effect on ribosomal RNA synthesis for the anthracycline antibiotic daunomycin in HeLa cells. Heterogeneous RNA was only partially inhibited and had altered physical characteristics. A continous synthesis of messenger RNA over a relatively long period was observed under conditions where ribosomal RNA synthesis was almost completely blocked. In yeast cells mitochondria are preferentially inhibited by the drug at a concentration of about 1o µg/ml in most strains. The replication of mitochondrial DNA is apparently unaffected, whereas mitochondrial RNA synthesis was primarily affected by daunomycin. In liver mitochondria RNA synthesis was inhibited about 3o% by 1o µg/ml and higher concentrations of the drug up to 2o µg/ml did not further depress RNA synthesis. It was suggested that daunomycin inhibits the synthesis of a particular fraction of liver mitochondrial RNA (EVANS et al., 1973).

3. Comparison with Ethidium Bromide

For the sake of comparison a few remarks should be made concerning the effect of ethidium bromide on the synthesis of nucleic acids.

NEWTON (1957) investigated the mode of action of this drug in the lower trypanosomid flagellate *Strigomonas oncopelti*. In ethidium bromide-treated cells, DNA synthesis is rapidly inhibited, whereas RNA and protein synthesis appear to be unaffected during the first few hours. A similar change in nucleic acid metabolism was found in ethidium bromide-treated bacteria by TOMCHICK and MANDEL (1964). The trypanosomocidal properties of ethidium bromide may be based on the specific action of the drug on kinetoplast DNA. Trypanosomes have a single but elaborate mitochondrion. The mitochondrial DNA is probably entirely localized in the kinetoplast, which appears as a specialized part of this mitochondrion. Evidence is presented to show that ethidium bromide induces the loss of kinetoplastic DNA in growing trypanosomatidae, but not in resting cells; ethidium bromide therefore seems to inhibit the mitochondrial DNA replication system specifically, probably in the same way as was shown for acriflavine (STEINERT et al., 1969). These authors studied the effects of ethidium bromide on ^{3}H-thymidine and ^{3}H-uridine incorporation by *Crithidia luciliae*, using autoradiography. The drug, at a concentration of 2o µg/ml, inhibits kinetoplast DNA synthesis up to 9o% after a few minutes. Nuclear DNA synthesis is also very rapidly affected but the inhibition does not exceed 28% under the same experimental conditions. Ethidium bromide inhibits ^{3}H-uridine incorporation into kinetoplast- and nuclear RNA, but this effect on RNA synthesis is slower than the very rapid process which affects DNA replication.

A preferential inhibitory effect of ethidium bromide on mitochondrial versus nuclear DNA synthesis was observed in cultured L cells (mouse) and BHK cells (Hamster) (NASS, 1972). In contrast nuclear DNA synthesis of L-cells, BHK and polyoma virus transformed BHK cells was actually stimulated. In all three cells lines the drug caused a structural alteration of covalently closed mitochondrial DNA. These changes were consisted of an increased degree of supercoiling and in addition. breakage of circular DNA without re-closing.

3. Particular Activities in Biological Systems

a) Viruses and Phages

The influence of anthracyclines on nucleic acid synthesis raises the question of their capacity to affect the replication of DNA and/or RNA viruses. The antiviral effect of daunomycin was investigated in detail by COHEN et al. (1969). Daunomycin was added to HeLa cell cultures at various times before and after infection with DNA viruses, e.g. herpes virus hominis or vaccina virus. The daunomycin-treated cells did not exhibit any specific cytopathogenic effect, thus indicating that daunomycin markedly inhibited the multiplication of the viruses. When added 3 or 4 hours after virus infection, daunomycin did not influence viral replication.

The effect of daunomycin on some representative RNA viruses, e.g. poliomyelitis and influenza, was also studied in HeLa cell cultures at various times before and after infection. There was a very significant degree of polio virus replication although there was some reduction of the total virus yield in the presence of daunomycin. The authors attribute this fall in titer to the reduced metabolic activity of cells subjected to the toxic action of daunomycin. Influenza virus is also resistant to the effect of daunomycin. Thus DNA viruses can be effectively inhibited by daunomycin while there is virtually no effect on the replication of RNA viruses such as poliomyelitis or influenza. It seems likely that the drug acts chiefly on DNA and DNA-dependent synthesis of RNA.

The effect of daunomycin on DNA and RNA phages was studied by PARISI and SOLLER (1964). The studies showed that daunomycin exhibits inhibitory activity towards DNA phages, e.g. *E. coli* B phages T1, T3, T4 and T6, but not towards RNA phages, e.g. *E. coli* K12 (H.frR$_1$) male specific RNA phage U$_2$. Such inhibitory activity does not effect adsorption, injection and lysis, but rather the growth cycle of the sensitive phages. The burst size is reduced to an extent which is roughly proportional to the concentration of daunomycin, but the time of the burst is not comparably affected. Daunomycin has a pronounced inhibitory effect during the latent period. The amount of DNA synthesized in daunomycin-treated infected bacteria was found to be significantly lower than in comparable controls. From these findings it was suggested that the main site of action of daunomycin within the cell is DNA - both as the infecting phage genome and the newly produced viral DNA.

b) Mutagenicity of Intercalating Dyes

While the intercalating drugs, the aminoacridines, are strong mutagens and are assumed to cause mutations by intercalation into the DNA double helix (for review see WARING, 1966, 1968), very little is known concerning the mutagenicity of the anthracyclines. TABACZYNSKI et al. (1965) were not able to find any mutagenic effect with anthracyclines in two bacterial systems. OSTERTAG and KERSTEN (1965) observed chromosome breaks in cultured human leukocytes treated with actinomycin and proflavine but not with anthracyclines. If one assumes that the anthracyclines intercalate between the base pairs of the DNA double helix *in vivo* as *in vitro*, and if one further assumes that frame shift mutations caused by aminoacridines are the consequence of intercalation, then it is extremely difficult to understand why the anthracyclines behave so very differently with respect to mutagenicity than do the aminoacridines.

c) Chemotherapeutic Aspects

A variety of cells in culture, including Schmidt-Ruppin sarcoma, Burkitt lymphoma and fourteen samples of human leukaemia were treated with daunorubicin or alternatively with the daunorubicin-DNA complex (TROUET et al., 1972). Under the fluorescence microscope the cytoplasm of cells exposed to the free drug was seen to become rapidly and diffusely fluorescent. In contrast, cells treated with the complex displayed a slow and progressive appearence of fluorescence clearly localized to cytoplasmic particles. Free daunorubicin diffuses rapidly across the cell membrane, invades the cytoplasm and enters the nucleus where, following combination with DNA, fluorescence is quenched and DNA replication and transcription are blocked. In contrast, the non-permeant DNA complex is taken up only by pinocytosis and accumulates within the lysosomes, still in quenched form. After digestion of the DNA by lysosomal DNAase fluorescence appears first in the lysosomes and later in the cytoplasm, tracing the pathway followed by the drug in its migration towards the nucleus. If the outside concentration of daunorubicin is sufficiently high the pinocytic mode of entry should be slower than direct diffusion and the cytotoxic effect retarded as was often observed. The DNA-daunorubicin, when administered by a single or by repeated intraperitoneal injections to either NMRI or DBA/2 mice was less toxic than free daunorubicin. Although the principle to use DNA as a carrier for chemotherapeutic purpuses is an interesting aspect, DNA may prove undesirable as carrier owing to its informational content, antigenicity and ability to induce plasma DNAase.

Another important aspect for the usage of anthrycyclines as antitumor drugs is the observation, that prolonged treatment with the antibiotics causes tumor cells to become insensitive for the drugs (DANØ, 1972).
By rat liver or rat kidney extracts in the presence of NADPH daunomycin and adriamycin are converted to daunomycinol or adriamycinol respectively. The enzyme, responsible for the conversion, daunorubicin reductase, was purified to homogeneity. Its physicochemical characteristics, kinetics and its amino acid composition was determined by FELSTED et al. (1974). The molecular weight is about 4o ooo. The conversion of daunomycin to daunorubicinol is nearly complete and daunorubicinol is the major excretion form of the drug and an important component in the action of daunorubicin.

C. Chromomycin, Olivomycin and Mithramycin

1. Origin, Biological and Chemical Properties

Chromomycin, olivomycin and mithramycin are chemically closely related antibiotics. Chromomycin is produced by *Streptomyces griseus* as a complex of compounds from which the main substance, chromomycin A_3, was obtained by TATSUOKA et al. (1958). Olivomycin is a product of *Streptomyces olivoreticuli* and was isolated by GAUSE et al. (1962). The principal component of the olivomycin complex was designated olivomycin I (BRAZHNIKOVA et al., 1962, 1964 a, b). Mithramycin was isolated from *Streptomyces sp.* by RAO et al. (1962).

All three compounds are very similar in their spectrum of antibacterial action. They inhibit the growth of gram-positive microorganisms such as *Staphylococcus aureus, B. subtilis* and *Mycobacterium smegmatis*, but are inactive against gram-negative bacteria and fungi. The antibiotics are also active against mammalian cells in culture and have been shown to possess carcinostatic properties (GAUSE and LOSHKAREVA, 1965; GAUSE, 1967). On acid hydrolysis chromomycin A_3 results in a lipid soluble fraction which contains the chromophore chromocinone and a water soluble fraction from which four different sugars have been isolated - the chromoses A, B, C, and D (MIYAMOTO et al., 1964 a; TATSUOKA et al., 1964). The detailed structures of the sugar components were described by MIYAMOTO et al. (1964 b, c, 1966). In chromomycin A_3 the chromoses A and D in this order are attached by a glycosidic bridge to the 6-position of the aromatic ring system of the chromophore and chromoses C and B are linked to the oxygen in position 2 of the non-aromatic ring (Fig. 45). The structure of olivomycin was elucidated by BERLIN et al. (1966). Olivomycin contains the aglycone olivin and four sugar components, isobutyrylolivomycose,

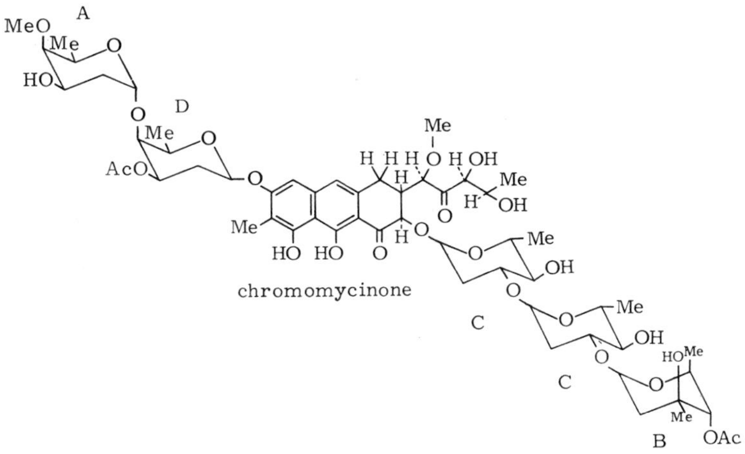

A —— D —— CHR —— C —— C —— B: chromomycin A_3 (AD-CHR-CCB)

derivatives:

D —— CHR —— C —— C —— B: D-CHR-CCB

D —— CHR —— C —— C : D-CHR-CC

D —— CHR : D-CHR

Fig. 45. Chromomycin and derivatives. (From: Inhibitors, Tools in Cell Research, Mosbach Colloquium 1969. Berlin-Heidelberg-New York: Springer 1969, p. 398)

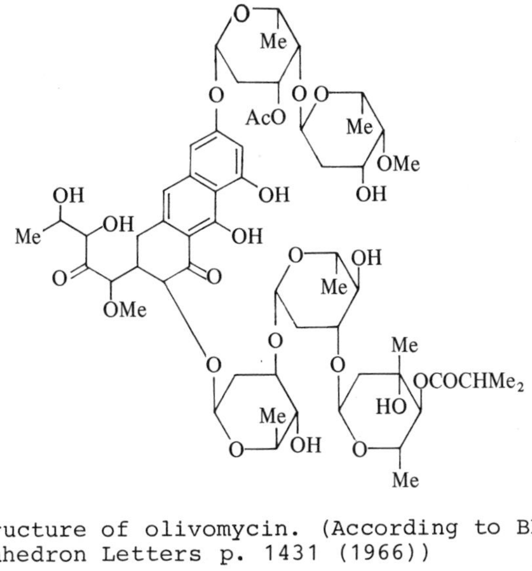

Fig. 46a. The structure of olivomycin. (According to BERLIN et al., 1966. From: Tetrahedron Letters p. 1431 (1966))

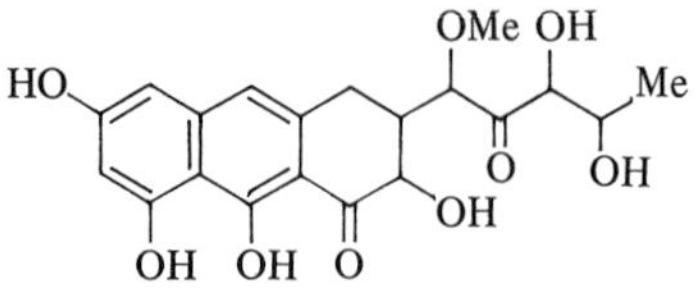

II
(olivin)

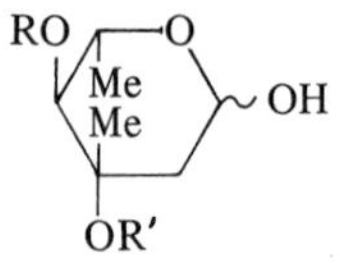

III: R = R′ = H
(olivomycose)
IV: R = Me_2CHCO, R′ = H
V: R = Me_2CHCO, R′ = Bz

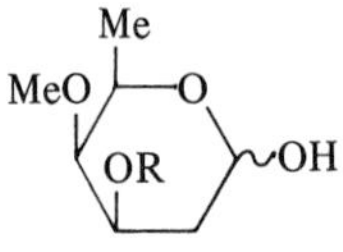

VI: R = H
(olivomose)
VII: R = Bz
VIII: R = Ac

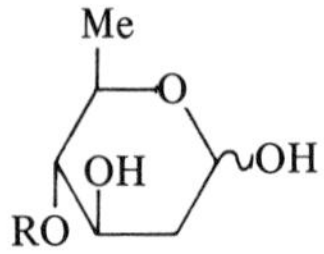

IX: R = H
(olivose)
X: R = Bz
XI: R = Ac

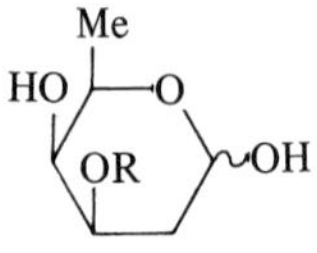

XII: R = H
(oliose)
XIII: R = Ac

Fig. 46b. The structures of olivin, olivomycose, olivomose, olivose and oliose. (According to BERLIN et al., 1966. From: Tetrahedron Letters p. 1433 (1966))

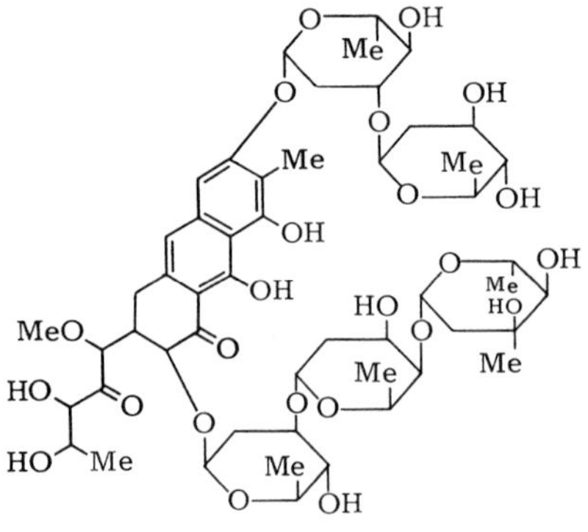

Mithramycin (= Aureolic acid) [11, 48]

Fig. 47a. The structure of mithramycin. (From: Inhibitors, Tools in Cell Research, Mosbach Colloquium 1969. Berlin-Heidelberg-New York: Springer 1969, p. 4o6)

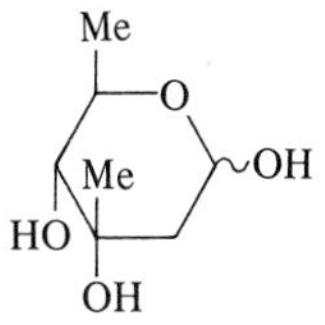

Fig. 47b. The structure of D-mycarose (*II*), D-olivose (*III*), D-oliose (*IV*), and chromomycinone (*V*). (According to BAKHAEVA et al., 1968. From: Tetrahedron Letters p. 3597 (1968))

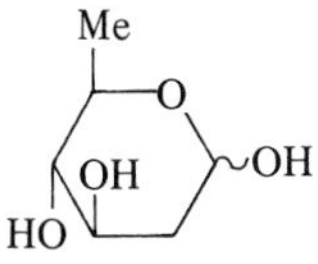

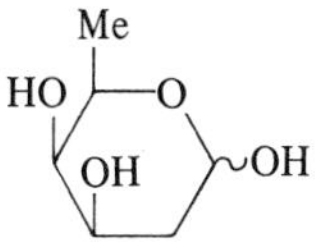

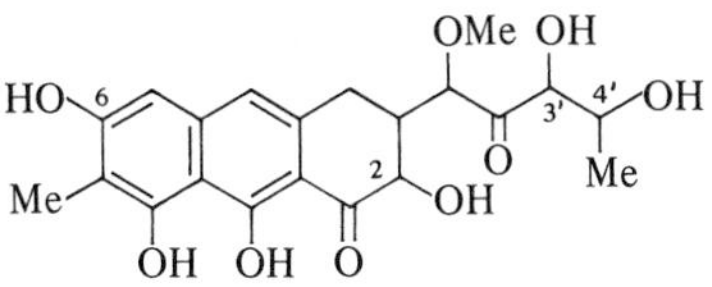

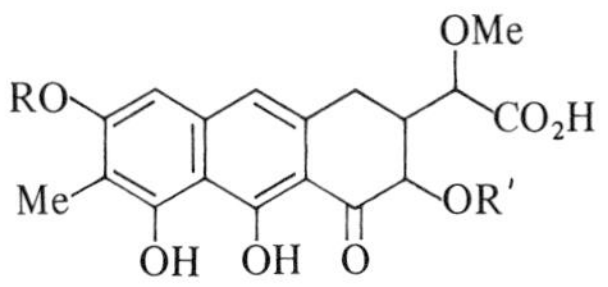

olivomose, olivose and oliose. The structures of the chromophore oli- vin, of the four sugars and of olivomycin are shown in Figure 46 a, b. The chromophore of mithramycin (aureolic acid) is identical to that of chromomycin A$_3$. The final structure of mithramycin including the five sugar residues was elucidated by BAKHAEVA et al. (1968). On acid hydrolysis, mithramycin yielded chromomycinone, D-mycarose, D- olivose, D-oliose in the ratio 1:1:3:1. Mithramycin contained two un- branched carbohydrate chains attached to the aglycone. Mithramycin is 2-[ß-mycarosyl (1→4)α-oliosyl (1→3)ß-olivosyl]-6-[ß-olivosyl (1→3)ß- olivosyl]-chromomycinone (Fig. 47 a,b).

On the basis of the close similarity in the UV spectra, other anti- biotics of as yet unidentified structure are closely related to the chromomycins. These are aburamycin, LA-7o17, M5 18 9o3, NSCA-649 (for summarizing references see MIYAMOTO et al., 1966).

2. On the Molecular Mechanism of Action

a) Interaction with DNA *in Vitro*

The chromomycin-like antibiotics appear to have essentially identical mechanisms of action; all of them form stable complexes with DNA *in vitro* (KERSTEN, 1968). Direct interaction of chromomycin with DNA was shown by its cosedimentation with DNA (KERSTEN and KERSTEN, 1965), and by a change in the absorption spectrum of the antibiotic on addition of DNA in the presence of Mg^{2+} (BEHR and HARTMANN, 1965; WARD et al., 1965). Olivomycin forms a stable complex with DNA which can be preci- pitated with ethanol (GAUSE et al., 1965).

The spectrum of chromomycin is shifted to longer wavelengths by DNA only in the presence of Mg^{2+} or other divalent cations which are es-

sential for complex formation. The UV parts of the spectra of chromomycin and mithramycin are decreased by native DNA also in the absence
of added Mg^{2+}, thus indicating that probably some orientation of the
antibiotic molecules along the DNA occurs in the absence of added
Mg^{2+} (KERSTEN and KERSTEN, 1968). From comparative studies with several DNA complexing dyes KERSTEN et al. (1966) concluded that chromomycin-like antibiotics differ from actinomycin and the anthracyclines
in their interaction with DNA. Chromomycin, mithramycin and olivomycin, while they cosediment with DNA they do not increase the sedimentation coefficient as do the anthracyclines. The chromomycin-like
antibiotics do not influence the viscosity of DNA solutions, nor do
they affect the melting temperature of DNA up to molar ratios of drug
to DNA-nucleotide of about 1:5. However at molar ratios of drug to
nucleotides of 1:1 and 5:1, chromomycin and mithramycin were found to
increase the melting temperature of DNA in the presence of Mg^{2+}, suggesting that these ligands are potentially able to stabilize the secondary structure of DNA (KAZIRO and KAMIYAMA, 1965; KERSTEN and KER-
STEN, 1968). The chromomycin-like antibiotics cause buoyant density
shifts of native DNA. The addition of Mg^{2+} ions is not necessary for
the density shifts to occur. The high concentration of Cs^+ during
density gradient centrifugation substitutes for Mg^{2+}.

To determine the relationship between the antibiotic-effected buoyant-
density shift of the DNA and its base composition, DNAs from various
bacterial species, varying in G+C content from 34 to 71%, and synthetic poly dAT were used. As in the case of the anthracyclines, the density shifts in the presence of the antibiotics tend to increase at
higher G+C content of the DNA, with relationships varying from approximately linear to highly irregular. No density shift of poly dAT was
observed with the chromomycin-like antibiotics, thus indicating that
G or C or both must be present in DNA for the interaction to occur.

The base specificity of complex formation between chromomycin and DNA
from different sources was studied in detail by BEHR et al. (1969).
The spectral changes occurring in chromomycin upon addition of DNA
were used to determine quantitatively the amount of bound chromomycin
in a mixture of DNA and antibiotic. In the presence of an excess of
inhibitor the apparent number of binding sites per base pair of the
polynucleotide is obtained. A Scatchard plot of binding isotherms
leads to comparable results. In addition this method gives the apparent binding constant K_{ap} for the equilibrium. The number of binding
sites depends on the source of DNA and rises with increasing guanine
content of the polynucleotide. While poly dAT does not interact at
all with chromomycin, the binding capacity of natural double-stranded
DNA increases linearly between 18 and 24 mole% guanine. It reaches a
limit at about 33 mole% guanine where about 1 molecule of chromomycin
is bound per 4 base pairs. DNAs with higher guanine content, such as
M. lysodeicticus DNA, exhibit no higher binding capacity. It is postulated that steric factors are responsible for the limitation of the
number of binding sites. The large, relatively lipophilic D-deoxy-
sugar side chains may be located preferentially in the grooves of the
DNA helix as a result of hydrophobic interactions. The sugar residue
probably occupies 3-4 nucleotide base pairs - as can be seen from
models.

Spectrophotometric titrations showed that heat-denatured calf thymus
DNA contains only o.o7 apparent binding sites per base pair compared
with o.19 of native DNA. Using equilibrium dialysis, KERSTEN and
KERSTEN (1968) came to similar conclusions. Interaction of chromomycin-like antibiotics with denatured DNA has also been demonstrated by
shifts in the buoyant density of dDNA on addition of the antibiotics

(KERSTEN et al., 1966). Heat-denatured DNA may still contain considerable portions of base paired regions. Therefore, these measurements
do not prove that base pairing is not required for complex formation.
BEHR et al. (1969) used poly (dG) · poly (dC) and its single-stranded
components poly (dC) and poly (dG) in their binding studies. None of
the single-stranded components interacts with chromomycin. Thus the
formation of a complex between chromomycin and DNA depends on base
pairing of the polydeoxynucleotides.

To elucidate more precisely the structural requirement for association with respect to the antibiotic molecule, BEHR et al. (1969) investigated interaction with chromomycins differing in the sugar side
chains. The association of chromomycins with DNA was found to be a
rather slow process. Chromomycin derivatives which contain fewer chromose residues than chromomycin do not differ in their rate of complex
formation with DNA. Thus the bulky side chains do not control the rate
of the association reaction responsible for the spectral shift. When
the dissociation rate was tested in the presence of 1% dodecylsulfate,
chromomycin separated completely from DNA during a 12-hour dialysis.
The rate of complex formation has been found to be independent of the
size of the sugar side chains; the rate of dissociation depends on
the number of sugars attached to the chromophore. It is postulated
that the dissociation rate of the complex determines the degree of
inhibitory potency.

What binding forces are involved in the association process has not
been solved as yet. By complex formation chromomycin protects guanine
segments of DNA very efficiently against the attack of nucleases. After enzymatic hydrolysis, chromomycin containing base-paired oligonu-
cleotide fractions remains in the reaction mixture. The antibiotic is
not covalently bound to these fragments (BEHR et al., 1969).

Although the complexes of chromomycin-like antibiotics with deoxyribonucleic acids have not been described in terms of molecular models,
these complexes can be discriminated from the actinomycin-DNA complexes and the anthracycline-DNA complexes by their physicochemical
properties. In contrast to the marked conformational changes associated with "intercalative binding", chromomycin-like antibiotics seem
to have much less influence on the conformation of DNA. Therefore,
it was postulated that chromomycin and mithramycin do not intercalate
(KERSTEN et al., 1966). In agreement with this concept are the results of WARING (197o), who studied whether the supercoils in closed
circular DNA change on binding of chromomycin or mithramycin (for
details see preceding section).

The mode of interaction of chromomycin A_3 with herring sperm DNA in
the presence of Mg^{2+} was studied by means of optical rotatory dispersion spectra (HAYASAKA and INOUE, 1969). Chromomycin A_3 exhibits
anomalous optical rotatory dispersion with multiple Cotton effects.
The addition of Mg^{2+} at lower levels resulted in a slight increase
in the rotatory power of the longest wavelengths Cotton effects,while
further addition of Mg^{2+} to the solution caused a marked decrease of
the Cotton effects. From these and other spectral data it was concluded that in the presence of Mg^{2+}, chromomycin A_3 in aqueous solutions exists as an equilibrium mixture of monomers and aggregates of
various sizes. The authors suggest that because of this aggregative
tendency, the antibiotic molecules bind to DNA to form an aggregative
state through the bridging with Mg^{2+} ions. In this state chromomycin
A_3 molecules are believed to bind preferentially to a site adjacent
to those already occupied in an antibiotic-DNA complex.

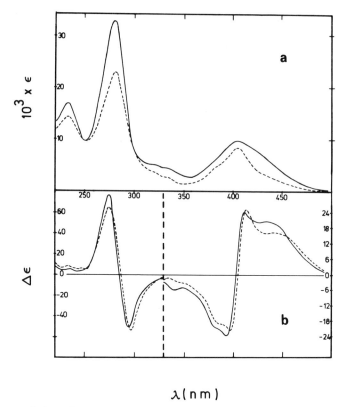

$$\lambda(nm)$$

Fig. 48 a and b. Absorption spectrum (*a*) and CD spectrum (*b*) of mithramycin ——— and chromomycin ----- in 1 x 1o^{-4}M phosphate buffer. (According to FEY, 1973)

Circular dichroism spectra of mithramycin- and chromomycin-DNA complexes were studied in detail (FEY and KERSTEN, 197o; KERSTEN, 1971; FEY, 1973). Mithramycin and chromomycin have characteristic intensive absorption bands at 4o5 nm and 28o nm (Fig. 48 a). In the CD spectrum two almost symmetrical doublets correspond to these absorption bands. On both sides of the doublets there are additional maxima at 445 nm and at 38o nm respectively (Fig. 48 b). The two doublets in the CD spectra of the antibiotics indicate that in a buffer solution at pH 7 and at a concentration of 1 x 1o^{-4} M, chromomycin or mithramycin forms dimers. The association occurs at the chromophores. The CD spectra of the antibiotics are changed quantitatively but not qualitatively on addition of Mg^{2+}. The amplitudes decrease, but the bands are not shifted to other wavelengths. The interaction of the antibiotics with DNA in the presence of Mg^{2+} is accompanied by drastic changes in the CD spectra, depending on the ratio of DNA-P to antibiotic (Fig. 49). On interaction with DNA at high ratios of DNA-P to drug (24:1), the doublets disappear, indicating that mithramycin or chromomycin is bound to DNA as a monomer. The CD spectra of the complexes of chromomycin with DNAs of different G+C contents revealed that the amount of drug which associated with DNA increased with increasing GC content of the DNA. The UV part of the CD spectra of the DNA-drug complexes are com-

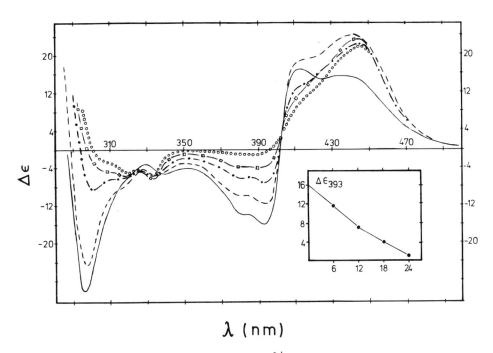

λ (nm)

Fig. 49. CD spectra of mithramycin-Mg^{2+}-DNA complex. Mithramycin
1×10^{-4}M. Molar concentration ratios mithramycin: Mg^{2+}: DNA = 1 :
0 : 24 ——— ; 1 : 2 : 6 ----; 1 : 2 : 12 -.-.-; 1 : 2 : 18-□-□- ;
1 : 2 : 24 oooo. The inserted diagram shows the decrease of the
amplitude at 393 nm caused by increasing the concentrations of DNA.
(According to FEY, 1973)

plicated. The easiest spectra to consider are those of the chromomycin-DNA complex. In a standard technique the CD spectrum of the DNA
within the complex was substracted from the CD spectrum of the complex
itself. A subsequent matrix rank analysis of the CD spectrum of the
DNA-chromomycin complex reveals that only free DNA unaltered by the
drug and one DNA-antibiotic complex were present. No cooperative structural effects on the DNA (as were observed with the anthracycline-DNA
complexes) occur: The DNA conformation is only slightly changed during the binding process. Mithramycin behaves similarly to chromomycin. The results indicate that mithramycin and chromomycin bind on
the outside of the DNA helix in GC-rich regions, thereby changing the
geometry of the DNA molecule significantly, but not as drastically as
the intercalating dyes.

An attempt has been made by NAYAK et al. (1973) to resolve the problem
of the role of Mg^{2+} in the binding of chromomycin to DNA. Spectrophotometric investigation of DNA-chromomycin A_3 complexes under various pH
and concentrations of Mg^{2+} were carried out. The investigators conclude from their results that Mg^{2+} ions are directly involved in binding to counteract the electrostatic repulsion between the negative
charged phosphate group of DNA and anionic chromomycin A_3.

b) Effects on DNA-Dependent Processes *in Vitro*

HARTMANN et al. (1964) and KOSCHEL et al. (1966) studied the effect of
chromomycin on DNA-dependent synthesis of DNA and RNA *in vitro* and found
that chromomycin preferentially inhibits the synthesis of RNA. In accordance with the findings that the chromomycin-like antibiotics preferentially associate with DNA of high GC content, the antibiotics are
rendered more effective in the *in vitro* system by using templates with
high GC content. From the detailed studies on the interaction of chromomycin and chromomycin-like antibiotics with DNA, it is possible to
explain the preferential inhibitory effect of chromomycin-like antibiotics on *in vitro* transcription (HARTMANN et al., 197o; BEHR et al.,
1969). During the process of transcription the RNA polymerase moves
along the DNA template; this is impossible in the case of the DNA-
antibiotic complex. The enzyme is not able to circumvent the antibiotic, so that the process of polycondensation comes to a stop. MÜL-
LER and CROTHERS (1968) pointed out that substances which do not bind
covalently to DNA do not totally inhibit DNA-dependent processes.
When the complex dissociates the enzyme can begin its catalytic activity again. The inhibitor is therefore expected to have increasing
efficiency with a lowering of its dissociation constant. That this
actually is the case is evident from the work of KOSCHEL et al.
(1966) and HARTMANN et al. (197o). The loss of chromose residues in
chromomycin - which is equivalent to an increase in the dissociation
constant - leads to a decrease in the inhibitory activity for RNA
polymerase.

To inhibit the DNA-dependent synthesis of DNA with DNA polymerase I
by 5o%, a hundredfold higher concentration than for RNA synthesis is
necessary. The reduced effect on DNA synthesis *in vitro* is possibly
due to the fact that the DNA polymerase starts from a double-stranded end or from a nick in the DNA template and reads base sequences
from single-stranded regions of the DNA. Chromomycin however complexes only with double-stranded DNA. This means that chromomycin is only
able to inhibit DNA synthesis by association with the end of the double-stranded region in the DNA. Since the binding occurs statistical-
ly over the entire double helical region, a considerably higher concentration of the antibiotic is necessary to block the end and thereby the synthesis of DNA. HONIKEL and SANTO (1972) studied the inhibitory effect of chromomycin A_3, actinomycin C_3 and daunomycin on DNA
synthesis and PPi-exchange reactions catalyzed by DNA polymerase I
in vitro. Chromomycin A_3 and actinomycin C_3, the latter when consider-
ing interaction with strong-binding sites, bind only to double-stranded DNA. In conjunction with the observed inhibition of the PPi-exchange
reaction carried out in the presence of only one nucleoside triphosphate this indicates that the *in vitro* inhibitory effect affects either
the complex-forming step or the initiation step of the DNA polymerase reaction.

c) Effects on Nucleic Acid Metabolism

WAKISAKA et al. (1963) first observed a selective inhibition by chromomycin A_3 of the biosynthesis of RNA in cultures of mammalian cells
(rabbit bone marrow cells and leukemic human leukocytes). In *B. subtilis* chromomycin inhibits the synthesis of DNA and RNA almost equally well (KERSTEN et al., 1967). In agreement with this, RATAPONGS
(1969) was not able to find any preferential effect of chromomycin
A_3 on RNA synthesis in Ehrlich ascites tumor cells and suggested that
the cytostatic effect of chromomycin A_3 is caused by both blockage
of RNA and DNA synthesis. The fact that chromomycin A_3 affects the

synthesis of DNA to virtually the same extent as the synthesis of RNA, was also observed by KAMIYAMA and KAZIRO (1966). According to KIDA et al. (1966) chromomycin A_3 inhibits the incorporation of adenine and uracil into RNA more than that of guanine and cytosine. Separation of the RNA fraction by column chromatography on methylated albumin showed that the degree of inhibition was related to the molecular weight of the components. Inhibition was in the order of 23 S RNA > 16 S RNA>soluble RNA. Incorporation of labeled precursors into DNA was altered by chromomycin to the same extent as their incorporation into soluble RNA.

Although chromomycin and mithramycin are closely related, they have different effects on nucleic acid synthesis - at least in microorganisms. Mithramycin preferentially inhibits the synthesis of RNA (KERSTEN et al., 1967). Nucleic acids from mithramycin-treated *B. subtilis* were separated on MAK columns. Mithramycin preferentially inhibits the incorporation of ^{14}C-uracil into fraction of the soluble RNA, corresponding to the 5 S RNA, and into 23 S ribosomal RNA. Mithramycin also affects the incorporation of radioactive precursors into rapidly labeled RNA. In higher organisms YARBRO et al. (1966) reported inhibition of ^{32}P-incorporation into cellular RNA of mouse ascites tumor cells and mouse liver cells by mithramycin under conditions in which ^{32}P incorporation into DNA was unaffected. The biochemical effects of mithramycin on the synthesis of macromolecules was studied in detail by NORTHROP et al. (1969) in mouse embryonic cells, BHK-21, and Chang's human conjunctiva cells. Following treatment with mithramycin, RNA synthesis was decreased by 85% in mouse embryo- and BHK-21 cell cultures, and by 45% in human conjunctiva cell cultures, when compared with controls. The inhibition of RNA synthesis in BHK-21 cells was dose-dependent and encompassed all cellular types of RNA. Although mithramycin inhibited protein and DNA synthesis in BHK-21 cells, this action was not dose-related, which suggests that the primary site of action of this antibiotic was on RNA synthesis.

A number of investigators have shown that olivomycin preferentially inhibits RNA synthesis in bacterial and animal cells. LAIKO (1962) reported that at a concentration of o.3 µg/ml, olivomycin completely stops the synthesis of RNA in staphylococcal cells, whereas the formation of DNA and of protein is much less affected. GAUSE et al. (1965) observed that in cultures of *B. megaterium*, olivomycin inhibits preferentially the synthesis of RNA. Because of the impermeability of their cell walls, *E. coli* are resistant to chromomycin-like antibiotics. However, it was demonstrated that olivomycin strongly inhibits the incorporation of ^{14}C-uracil into the RNA of *E. coli* protoplasts.

ZALMANZON et al. (1965, 1966) reported that olivomycin selectively inhibits the synthesis of RNA of human amnion cells (strain FL) in tissue culture. GAUSE and LOSHKAREVA (1965) measured the incorporation of ^{32}P into DNA and RNA in Ehrlich ascites carcinoma of mice and also reported a preferential effect of this antibiotic on the synthesis of RNA.

In two bacterial systems in which proflavine was a strong mutagen, TABACZYNSKI et al. (1965) were unable to find any mutagenic effect of actinomycins, chromomycin-like antibiotics or anthracyclines. With human leukocytes in culture, OSTERTAG and KERSTEN (1965) observed that proflavine and actinomycin caused chromosome breaks. However neither mithramycin nor chromomycin caused chromosome breaks in this system. If one compares the effects on DNA of chromomycin-like antibiotics with those of actinomycin and anthracyclines, it is obvious, especially from the CD measurements, that the different classes of antibiotics cause changes in the conformation of DNA which in every case lead to a different DNA geometry. In addition this effect depends on the GC content of the DNA. Thus it is plausible that, with respect to mutagenicity, the antibiotics exhibit quite different effects.

3. Particular Activities against Viruses

SMITH and HENSON (1965) reported that mithramycin interferes with the multiplication of pseudorabies virus, but not with that of encephalo- myocarditis virus. SCHOLTISSEK et al. (197o) studied the action of mithramycin on the synthesis of fowl plaque virus (influenza A) and Newcastle disease virus (parainfluenza). They showed that mithra- mycin inhibited RNA synthesis in chick embryo cells in culture almost as efficiently as actinomycin D although the inhibition was consider- able delayed. There was no direct effect on cellular protein synthe- sis. Mithramycin interfered with the multiplication of fowl plague virus (influenza A) but had little effect on the multiplication of the Newcastle disease virus (parainfluenza). Mithramycin, when added immediately after infection, preferentially inhibited the synthesis of fowl plague minus strand RNA in culture, but had only a slight ef- fect on the production of plus strand RNA. The synthesis of virus RNA-dependent RNA polymerase and RNP antigen was only slightly inhi- bited, while the production of hemagglutinin and neuraminidase was strongly affected. Thus mithramycin, although it has no direct effect on cellular protein synthesis, interferes with the production of vi- rus-specific proteins. The synthesis of the virus components is not uniformly inhibited. Only a little hemagglutinin, much more neurami- nidase, and almost the normal amount of RNP antigen is produced. As pointed out by the authors, mithramycin may affect the synthesis of different parts of the viral RNA in different ways and intensities. Some of the viral proteins may be translated from the plus strand RNA and some from the minus strand RNA, and mithramycin may affect each strand differently.

D. Kanchanomycin

1. Origin, Biological and Chemical Properties

Kanchanomycin is produced by an as yet unidentified *Streptomyces* spe- cies. LIU et al. (1963) isolated kanchanomycin and showed its bac- terial and tumoricidal properties.

This antibiotic forms orange crystals, contains 61.66% C, 4.64% H, 5.3% N, and 5.67% O-Me. The structure of kanchanomycin is not yet known. The molecular weight is assumed to be 6oo.

2. On the Mechanism of Action

Kanchanomycin interacts *in vitro* with polynucleotides in the presence of Mg^{2+} ions (FRIEDMAN et al., 1969a). This antibiotic also interacts with DNA *in vitro*. Interaction with DNA was shown by spectral shifts of the dye molecule upon addition of DNA in the presence of Mg^{2+}. Two types of complexes are formed, named (I) and (II). The first type of

complex arises rapidly and is slowly converted into the second form
(II). Complex formation is completely inhibited by o.1 M NaCl. Complex (I) but not complex (II) can be partially dissociated by NaCl;
complex (I) is readily dissociated by EDTA, complex (II) only slowly
dissociated by EDTA. Kanchanomycin in the complex (II) decreases the
buoyant density of $d(A-T)_n$, increases the sedimentation rate of $d(A-T)_n$
and increases the viscosity of deoxyribonucleic acid. The spectral
changes require a Mg^{2+} concentration equal to the kanchanomycin concentration. The retention of the kanchanomycin-Mg^{2+}-polynucleotide
complex on Millipore filters and the changes in viscosity of DNA require a Mg^{2+} concentration equal to the nucleotide concentration.
The binding of kanchanomycin to polynucleotides appears to be cooper-
tive; there seems to be a greater attraction of a given kanchanomycin
molecule to a DNA molecule already binding one or more kanchanomycin
molecules, than to a DNA molecule with no bound kanchanomycin. Cooperative binding to polyribonucleotides has also been shown for acridine
orange (BEERS and ARMILEI, 1965).

FRIEDMAN et al. (1969b) employed optical rotatory disperion and circular dichroism measurement to study the binding of kanchanomycin to
deoxyribonucleic acid, ribonucleic acid and synthetic polynucleotides.
Free kanchanomycin has optical activity, which may arise either from
an inherent dissymmetry in the molecule or from polymerization. Since
the structure of kanchanomycin is currently unknown, it is difficult
to ascertain the source, from which its dissymmetry derives. The optical rotatory dispersion properties of kanchanomycin remain unchanged in 8 M urea, 2 M NaCl, or in 1oo% dimethylformamide, which
suggests that the antibiotic does not polymerize in aqueous solution
at the concentrations employed in the studies. Kanchanomycin exhibits
an extrinsic Cotton effect in the presence of Mg^{2+} and polynucleotides.
In the presence of Mg^{2+} the antibiotic forms two distinct complexes
with deoxyribonucleic acid, ribonucleic acid and certain other polynucleotides. Complex (I) dissociates readily and complex (II) slowly
on removal of Mg^{2+} with a chelating agent. Thus it is unlikely that
covalent bonds are formed in either complex. The initial complex between kanchanomycin, Mg^{2+} and polynucleotides seems to involve electrostatic interactions between the negatively-charged phosphate backbone of the polynucleotide and a positively charged kanchanomycin-
Mg^{2+} complex. The nature of the second complex is not precisely known,
but it probably involves the bases of the polynucleotide. For complex
I formation, there are as many binding sites per polynucleotide as
there are bases.

JOEL et al. (197o) studied the *in vitro* effect of kanchanomycin on
DNA-dependent RNA polymerase and DNA-dependent DNA polymerase. Of all
the reported antibiotic inhibitors of nucleic acid synthesis, kanchanomycin resembles luteoskyrin most closely with respect to its DNA
binding properties. Kanchanomycin inhibits both the synthesis of DNA
and the synthesis of RNA in the *in vitro* system. Inhibition by the
DNA-dependent DNA polymerase I can be overcome by increasing the concentration of the DNA polymerase. Increasing the concentration of the
DNA template relative to kanchanomycin does not overcome the inhibition of the DNA-dependent RNA polymerase reaction. It was concluded
that the inhibition of RNA synthesis is not caused only by the binding of the agent to the DNA, but must involve an inactivation of the
polymerase in the complex. When more enzyme is added, inhibition
caused by kanchanomycin is overcome, because the enzyme is able to
attach to initiation sites which are free of the inhibitor.

It is not clear whether kanchanomycin exerts its toxic effect on
cells primarily by direct interaction with nucleic acid synthesis
within the cell.

E. Distamycin and Netropsin

1. Origin, Biological and Chemical Properties

Distamycin A is produced by *Streptomyces distallicus*. The antibiotic is of special interest because of its preferential inhibitory effect on animal viruses (CASAZZA and GHIONE, 1964/1965; VERINI and GHIONE, 1964; WERNER et al., 1964; FOURNEL et al., 1965). The antibiotic exhibits antibacterial and antifungal activity and interferes with the development of T_1 and T_2 phages in *E. coli* (DI MARCO et al., 1963 a, b). Experimental tumors are inhibited by distamycin and by the closely related antibiotic netropsin (DI MARCO et al., 1962).

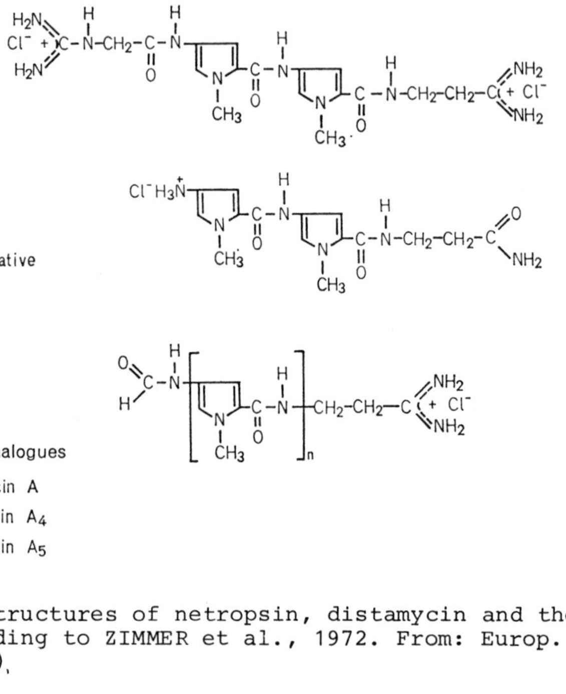

Netropsin

Netropsin derivative

Distamycin analogues

n = 3, Distamycin A

n = 4, Distamycin A_4

n = 5, Distamycin A_5

Fig. 5o. Chemical structures of netropsin, distamycin and their derivatives. (According to ZIMMER et al., 1972. From: Europ. J. Biochem. 26, 82 (1972))

Both distamycin and netropsin are basic oligopeptides. The structure of distamycin A, identified and confirmed by total synthesis by ARCAMONE et al. (1964), is characterized by three residues of 1-methyl-4-aminopyrrole-2-carboxylic acid and two side chains, the first being a formyl group and the second a propionamidine chain. ARCAMONE et al. (1969) succeeded in synthesizing some structural analogs of distamycin A. Structural modifications were obtained by substitution of the formyl group, substitution of the propionamidine side chain and variation of the number of pyrrole residues. The chemical structures of distamycin, netropsin and their derivatives are shown in Fig. 5o.

2. On the Molecular Mechanism of Action

a) Interaction with DNA *in Vitro*

Interaction of distamycin with DNA has been studied by several groups
of investigators. ZIMMER et al. (1971a, 1972) showed that the addition of
distamycin A or netropsin to DNA caused a decrease in the ultraviolet
absorption of DNA. The spectra of the DNA-antibiotic complexes are
hypochromic relative to DNA in the region between 22o and 3oo nm with
a new maximum appearing at 325 nm or 34o nm for the netropsin- and
distamycin A-DNA complexes respectively. At these wavelengths neither
DNA nor netropsin nor distamycin A shows absorption bands in its spectrum (Fig. 51 a, b). The hypochromicity depends markedly on the base
composition of DNA and is more pronounced for AT-rich DNA than for
GC-rich DNA.

The antibiotics shift the thermal transition of DNA to higher values;
the hyperchromicity is increased; the transition width - defined as
a temperature range ΔT in which the hyperchromicity increases - is
narrowed. The enhancement of the thermal stability of DNA is more pronounced in the presence of netropsin than in the presence of distamycin.

Optical rotatory dispersion and circular dichroism measurements (ZIM-
MER and LUCK, 197o) also indicated pronounced changes in the DNA conformation. In agreement with the optical transitions, 325 nm and 34o nm
Cotton effects were found in these wavelength regions. The authors
discuss whether the hypochromicity may reflect changes in the base
arrangements, and some orientation of the bound antibiotic. The appearance of ultraviolet absorption maxima at longer wavelengths is
also explained by changes in the arrangement of bases. They conclude
that the binding of these antibiotics is accompanied by an alteration
of the stacking interaction in double helical regions. Unstacking of
the bases at higher temperatures results in complete reduction of the
binding effects. Attachment of netropsin and distamycin A to denatured
DNA is thought to be caused by an affinity to the unspecific base-pair
regions contained in the secondary structure of denatured DNA at room
temperature.

ZIMMER and LUCK (1972) found that urea together with 2 M LiCl completely disorganizes the complex between DNA and netropsin, whereas the
distamycin DNA complex can still be formed under these conditions. It
is considered that hydrophobic interactions, hydrogen bonding, dispersion forces, besides electrostatic interactions, maintaining the
structure of the DNA oligopeptide complexes.

ZIMMER et al. (1972) extended their studies to analogs of distamycin
and netropsin with the following additional results: an increasing
number of methyl-pyrrole groups from 3 (occurring in distamycin A) to
5 (Fig. 5o) enhances the thermal stability of the complex with DNA.
(This was also observed by CHANDRA et al. (1971).) Elimination of the
basic groups of the netropsin molecule decreases drastically the melting temperature of the oligopeptide-DNA complex.

Circular dichroism spectra of the complexes (Fig. 52) exhibit a pronounced band at the longer wavelengths - as is to be expected from
the UV spectra. The induced circular dichroism in the absorption region of the oligopeptide is interpreted in terms of the appearance
of chirality in the bound netropsin molecule. The results indicate
that chirality increases in the complex with increasing number of chro-

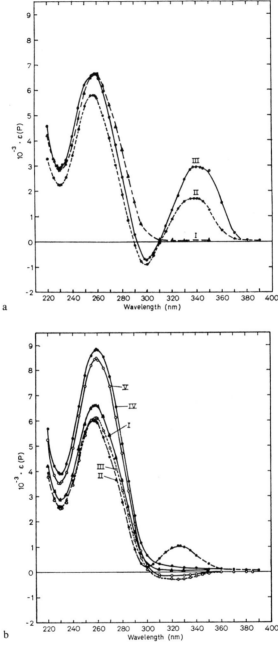

Fig. 51 a and b. *a*) Ultraviolet absorption spectra of the DNA.distamycin complexes in 2o mM NaClO₄, at o.2 mol oligopeptide/DNA-P; (*I*) DNA; (*II*) DNA + distamycin A; (*III*) DNA + distamycin A₄. (According to ZIMMER et al., 1972. From Europ. J. Biochem. <u>26</u>, 83 (1972))
b) Ultraviolet absorption spectra of the DNA.netropsin derivative complex in 2o mM NaClO₄ at o.2 mol oligopeptide DNA-P; (*I*) DNA; (*II*) DNA + netropsin; (*III*) DNA + netropsin derivative; (*IV*, *V*) after heating of (*I*) and (*II*) to 95°C. (According to ZIMMER et al., 1972. From: Europ. J. Biochem. <u>26</u>, 84 (1972))

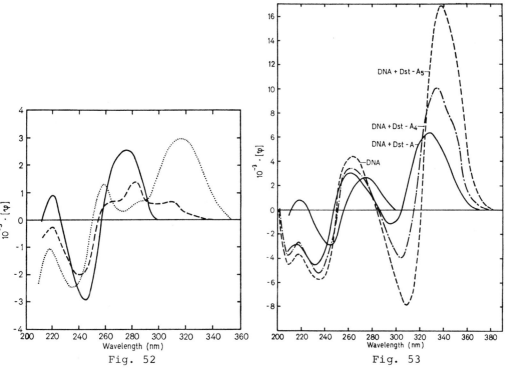

Fig. 52

Fig. 53

Fig. 52. Circular dichroism spectra of calf thymus DNA on interaction
with netropsin and its derivative in o.o2 M NaClO4, pH ≈ 6.5, at≈o.1
mol oligopeptide/DNA-P. ————, DNA; ...., DNA.netropsin; ----,
DNA.netropsin derivative. (According to ZIMMER et al., 1972. From:
Eur. J. Biochem. 26, 86 (1972))

Fig. 53. Circular dichroism spectra of calf thymus DNA on interaction
with distamycin A and its analogs in o.o2 M NaClO4, pH ≈ 6.5 at≈o.1
mol oligopeptide/DNA-P. (According to ZIMMER et al., 1972. From:
Eur. J. Biochem. 26, 87 (1972))

mophores of the oligopeptides (Fig. 53). The authors conclude that e-
lectrostatic attraction of the oligopeptides by interaction of the pro-
pionamidino- and guanidino groups with the negative phosphate sites
of DNA, is important for the formation of the complex. KREY and HAHN
(197o) studied the complex of distamycin A with calf thymus DNA by
using light absorption, optical rotatory dispersion (ORD spectra)
and flow dichroism. The absorption maximum of distamycin at about
3oo nm is shifted to higher wavelengths on addition of DNA. Spectrophotometrically it was shown that distamycin A displaced methyl green
from its DNA complex, indicating that distamycin competes with methyl
green for the same binding sites. The flow dichroism of the purine/
pyrimidine and of distamycin in the DNA-antibiotic complex was of the
same magnitude but of opposite sign. From these data it is suggested
that the N-methylpyrroles of distamycin are placed in an orderly ar-
ray relative to planes of the base pairs of the DNA in the complex.

The complex of distamycin with DNA is relatively more stable to urea
and inorganic ions than many other DNA-ligand complexes (HAHN et al.,

1966; ESTENSEN et al., 1969).KREY and HAHN (197o) suggest that the propionamidine residue of the distamycin molecule is electrostatically attracted to the phosphate groups of DNA. Other forces must also be involved in the formation of the complex. The positive DNA-induced Cotton effect in the ORD spectrum of distamycin is possibly caused by conformational changes arising from N-alignment of the N-methylpyrroles with the helix. That Cotton effects are also induced by a single-stranded DNA agrees well with the findings that distamycin forms complexes with denatured DNA which melt cooperatively like a highly ordered structure (CHANDRA et al., 197o). From the changes of the UV spectrum of distamycin-DNA complexes and the chromatographic behavior of distamycin on a DNA cellulose column (for details see chapter on anthracyclines), ZUNINO and DI MARCO (1972) assumed the existence of two classes of binding sites in DNA for the antibiotic. ZIMMER et al. (1971 b) presented evidence to show that two types of binding sites exist in DNA for netropsin.

Viscosity measurements by REINERT (1972) indicate that A-T clusters in DNA are the sites most responsible for netropsin induced conformational changes. Considerable local conformational changes of DNA with drastic alterations of the geometric parameters can be generated by suitable substances at DNA sites of special sequence. If a specific protein upon interaction with DNA results in changes of mutual distances between complementary groups of DNA and protein this would allow additional secondary bonds to be formed combined with additional gain in free energy and binding strength. Thus the studies of peptide antibiotic DNA interactions contribute valuable information on the problem of nucleic acid protein interactions.

b) Effect on DNA-Dependent Processes *in Vitro*

Distamycin A was found strongly to inhibit the template activity of DNA for DNA polymerase I (PUSCHENDORF and GRUNICKE, 1969). In the DNA polymerase assay, a 1oo,ooo x g supernatant of homogenates from Ehrlich ascites tumor cells was used as a source of enzyme and calf-thymus DNA as substrate. The concentrations of distamycin needed to produce a significant inhibition of DNA template activity are in the same range as those required for its antiviral or antitumor effect. It is suggested that the antiviral and cytostatic activity of distamycin A is caused by its interaction with DNA.

CHANDRA et al. (197o) studied the effect of distamycin A on the structure and template activity of DNA in the RNA polymerase system. RNA polymerase was isolated from *E. coli* by means of the procedure of ZILLIG et al. (1966). Calf thymus DNA was used as template. Distamycin A inhibits the template functions of native as well as of denatured DNA for RNA polymerase. The template activity of native DNA is more sensitive to the antibiotic than that of denatured DNA. With the DNA-dependent DNA polymerase, PUSCHENDORF and GRUNICKE (1969) found that denatured DNA is more sensitive to distamycin A in the DNA polymerase reaction. ZIMMER et al. (1971 a) compared the effect of netropsin and distamycin A on DNA and RNA polymerase reactions. In the DNA polymerase system denatured DNA proved to be more sensitive to the action of these antibiotics than was native DNA. The reverse, e.g. a stronger inhibition in the presence of native DNA, was observed by using the RNA polymerase assay. PUSCHENDORF and GRUNICKE (1969) suggested that this difference is caused by the fact that DNA polymerase I prefers single-stranded DNA as template, whereas RNA polymerase is more active on double-stranded DNA. The inhibition of RNA synthesis

can be attributed entirely to an interaction of the antibiotic with
the DNA templates. According to the results of ZIMMER et al. (1971 a),
the inhibition of template activity should be attributed to an interference with AT-rich regions rather than with GC-rich sites in DNA,
despite the fact that the mechanism of interaction with DNA differs
for netropsin and distamycin.

PUSCHENDORF et al. (1971) studied the mechanism by which the DNA-dependent RNA polymerase system is blocked by distamycin A. The elongation of growing chains was found to be resistant to the action of the
drug up to 4 min after addition of the antibiotic. The influence of
distamycin on the formation of the enzyme-DNA complex was followed
by density gradient centrifugation. Distamycin A inhibits the binding
of RNA polymerase to DNA.

The existence of virion-associated DNA polymerase in oncogenic RNA
viruses (reversed transcriptases) led CHANDRA et al. (1972 a) to investigate the effect of distamycin and its derivatives on the DNA
polymerase activities of MSV-M (murine sarcoma virus) virions and
FLV (Friend leukemia virions) with various templates. The results
show that the inhibition of DNA polymerase activities of FLV and
MSV-M by distamycin is template specific. Templates containing thymine and adenine are highly sensitive to the action of distamycins.
By means of structural analogs of distamycin (Dist. A4 and Dist. A5,
Fig. 5o), it was shown that the antiviral activity increases (CHANDRA
et al., 1972 a, b) while the cytotoxic effect decreases with increasing numbers of chromophores. KOTLER and BECKER (1971) also observed
inhibition of reversed transcriptase in Rous sarcoma virions.

Distamycin and netropsin differ from the other DNA-complexing antibiotics in that, within a certain concentration range, they inhibit
the multiplication of DNA phages (T_1 and T_2) and the multiplication
of animal viruses (vaccinia, herpes simplex and adenovirus) while
the synthesis of host DNA or host RNA remains unaffected. Distamycin
and netropsin may become useful antibiotics against viral infections.

F. Anthramycin

1. Origin, Biological and Chemical Properties

Anthramycin (refuin) was isolated from the fermentation broth of *Streptomyces refuineus* var. *thermotolerans* and was found to be active against a
variety of experimental tumors (TENDLER and KORMAN, 1963). The active
component of refuin was identified as anthramycin (LEIMGRUBER et al.,
1965 b). The biological activity of anthramycin has been studied extensively by GRUNBERG et al. (1966) who observed cytotoxic activity
of the antibiotic *in vitro* against monkey kidney, HeLa and rabbit
kidney cells. Furthermore Ehrlich ascites cells, sarcoma 18o, Ehrlich
solid carcinomas, Walker carcinosarcoma, human epithelioma (No. 3)
and human adenoma (No. 1) are inhibited by anthramycin. The antibiotic also exhibits marked activity against a localized subcutaneous infection with *Trichomonas vaginalis*. Anthramycin exhibits a fairly wide
in vitro antimicrobial spectrum, being most active against *E. coli* B
infected with bacteriophage T4, gram-positive bacteria, mycobacteria
and streptomycetes, with weaker activity against fungi and yeast.

However, at tolerated doses it is inactive against any bacterial infections *in vivo*.

The structure and stereochemistry (Fig. 54) have been elucidated
(LEIMGRUBER et al., 1965 a) and the total synthesis of anthramycin was
accomplished by LEIMGRUBER et al. (1968). The chemical name is (5, 1o,
11, 11a-tetrahydro-9,11-dihydroxy-8-methyl-5-oxo-1H-pyrrolo[2,1-c]
[1,4]-benzo-diazepine-2-acrylamide). When anthramycin is crystallized
from a hot methanol-water mixture, anthramycin methyl ether (AME) is
formed. This substance is more stable than anthramycin and, in addition, exhibits antitumor and chemotherapeutic activity equal to that
of the parent antibiotic. Nuclear magnetic resonance spectroscopy
showed that AME in aqueous solution was converted to anthramycin. The
conversion may be followed by a slight change in the absorption spectrum. At pH 7 and 37°C the change is complete within 1 h (KOHN and
SPEARS, 197o).

2. On the Molecular Mechanism of Action

a) Interaction with DNA *in Vitro*

Interaction of anthramycin with DNA was studied by KOHN et al. (1968),
who observed that in L121o leukemia cells DNA and RNA synthesis were
inhibited preferentially. DNA was found to cause a gradual change in
the UV absorption spectrum of AME, producing a bathochromic shift of
the absorption maximum from 333 nm to 343 nm and a reduction in the
hight of the peak to 82% of that of the free anthramycin. The spectrophotometric changes and thus the reaction of anthramycin with DNA are
time-consuming. The reaction is reported to be completed within a period varying from seconds to 45 minutes (HORWITZ, 1971). Direct interaction of anthramycin with DNA was also shown spectrophotometrically
by STEFANOVIĆ (1968). Kinetic studies suggested that both the free
compound and the methylether can react with DNA, but at different rates. Anthramycin reacts much faster than its methyl ether. One molecule of anthramycin is bound for approximately each eight base pairs
of calf thymus DNA. Anthramycin displaces methyl green from a DNA-
methyl green complex and is a competitive inhibitor (with respect to
DNA) in the enzymic hydrolysis of DNA by DNAase I (BATES et al.,
1969).

The melting temperature of DNA is increased on interaction with the
antibiotic in proportion to the molar ratio of antibiotic to DNA nucleotide. The melting profiles are not reversible on brief cooling,
indicating that the complexes do not contain stable interstrand cross-
links. The complex of AME with *E. coli* DNA is characterized by a reduced buoyant density in CsCl. When AME is reacted with denatured DNA,
only a small decrease in buoyant density occurs - even after a relatively long reaction time. It appears that although anthramycin reacts only weakly or slowly with denatured DNA, it remains bound after
denaturation of the native DNA-anthramycin complex (KOHN et al.,
1968).

The anthramycin-DNA complex is remarkably stable. It survives alcohol
precipitation, high salt concentration, temperatures up to the T_m and
alkali denaturation of the DNA. Anthramycin failed to show any interaction with deoxyribonucleotides. KOHN and SPEARS (197o) discuss the
nature of the anthramycin-DNA interaction and the specificity of anthramycin for DNA. Compared with most of the known antibiotics that
bind firmly but not covalent to DNA, anthramycin is relatively small.

The molecule lacks side chains which are essential for the tight binding of actinomycin, chromomycin-like antibiotics and anthracyclines,
and has no positive charge, as daunomycin, nogalamycin and cinerubine. Anthramycin is not planar as is ethidium bromide. The molecule
lacks all of the features that have been associated with tight non-
covalent binding to DNA. KOHN and SPEARS (1970) studied molecular
models and did not find any obvious way to fit the anthramycin molecule into the structure of the DNA helix. Yet the anthramycin-DNA bond
is remarkably strong and the reaction shows specificity for helical
DNA. Some indications of the high stability are: the anthramycin-DNA
complex does not dissociate upon prolonged dialysis or repeated gel
filtration; anthramycin is scarcely dissociable from DNA upon repeated alcohol precipitation; sodium lauryl sulfate, which has been reported to dissociate the DNA complexes of actinomycin and chromomycin,
has no effect on the anthramycin-DNA complex; silver ions under conditions that dissociate the DNA complexes of actinomycin and daunomycin also have no effect on the anthramycin-DNA complex; when the
complex is denatured with alkali the single strands obtained contain
attached anthramycin. Anthramycin can be separated from the DNA complex by precipitation of the DNA with trichloracetic acid; the released compound is unchanged anthramycin. It is suggested that covalent
bonds may be involved in the formation of the complex.

The binding of anthramycin to DNA was found to protect anthramycin
against degradation by heat or high pH. KOHN et al. (1968) offer an
explanation for the protective effect: the anthramycin molecule may
have a reactive site that is responsible for both its degradation and
the binding to DNA. The increased stability of bound anthramycin might
then be due to a protection of this reactive site through its covalent
bonding to DNA.

The anthramycin-binding reaction appears to be specific for DNA-like
polymers and to be sensitive to base sequence and polymer conformation.
The specificity for deoxynucleotide polymers is shown by the absence
of any reaction with RNA or with deoxymononucleotides. Anthramycin
reacts with a dG:dC polymer, but not with a rG:rC polymer, nor with
the copolymer d(AT). The overall base composition of a DNA however
does not influence the binding capacity.

b) Inhibition of DNA-Dependent Processes *in Vitro* and within the Cell

Effects of anthramycin on RNA polymerase reaction and DNA polymerase
I reaction were studied by HORWITZ and GROLLMAN (1968) and by BATES
et al. (1969). Anthramycin was found to be a potent inhibitor of the
DNA-dependent RNA polymerase reaction of *E. coli*, inhibiting the *in
vitro* system by more than 50% at a concentration of $2 \times 10^{-6}M$. Under
identical conditions, actinomycin D ($2 \times 10^{-6}M$) inhibited the reaction by 80%. The degree of inhibition by anthramycin was greatly diminished when heat-denatured DNA was substituted for native calf thymus DNA as a primer. In contrast to other antibiotics anthramycin did
not inhibit RNA polymerase reaction with double-stranded synthetic
polynucleotides (poly dAT, poly dGdC) as primers.

Anthramycin inhibits the synthesis of DNA and RNA in HeLa cells by
90% at concentrations which have no effect on protein synthesis. Inhibition was complete within 10 min of adding anthramycin to the cultures (HORWITZ and GROLLMAN, 1968). Similar effects of anthramycin
on nucleic acid synthesis in L.1210 cells have been reported by KOHN
et al. (1968) and on nucleic acid synthesis in Ehrlich ascites cells
by BATES et al. (1969). A preferential inhibitory effect of anthra-

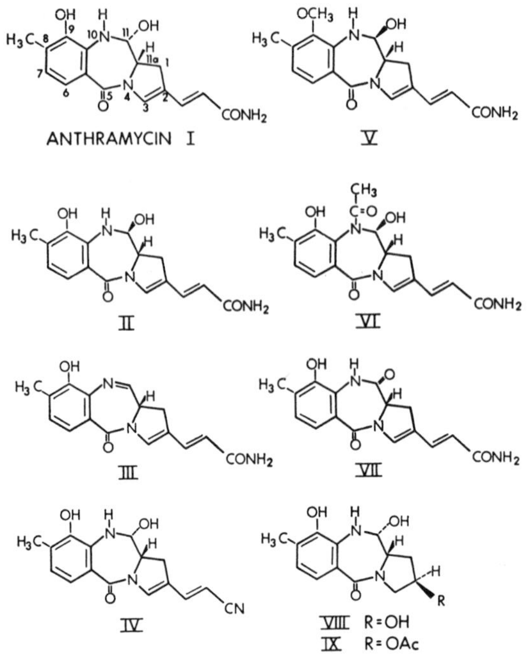

Fig. 54. Structural formulas of anthramycin and related derivatives. (According to LEIMGRUBER et al., 1965. From: Progr. Mol. and Subcell. Biology 2. Berlin-Heidelberg-New York: Springer 1971, p. 41)

mycin on RNA synthesis was observed in *E. coli* (HORWITZ and GROLLMAN, 1968).

c) Chemosterilant Action of Anthramycin

The activity of anthramycin and structurally-related analogs as chemo-sterilants of the housefly, *Musca domestica L.*, correlates closely with the action of these compounds as inhibitors of *E. coli* RNA polymerase. Since inhibition of RNA polymerase by anthramycin reflects the bind-ing of this antibiotic to the DNA primer, HORWITZ et al. (1971) sug-gest that the interaction of anthramycin with DNA may account for its action as a chemosterilant. The authors compared anthramycin and seven structurally-related analogs (Fig. 54) with respect to their chemo-sterilant activity and their effect on the DNA-dependent RNA polymer-ase reaction with the *E. coli* system. The results of these structure-activity studies indicate that the ability of anthramycin and deriva-tives to act as chemosterilants in houseflies correlates closely with the inhibitory effect of these compounds on the RNA polymerase of *E. coli*. Analogs in which the phenolic hydroxyl in position 9 is me-thylated, the aniline nitrogen is acetylated, the carbinol amine func-tion is replaced by an amide group, or in which the conjugated side chain is absent, do not exhibit the sterilizing effects of anthramy-cin. These compounds do not inhibit the RNA polymerase of *E. coli*, and it is concluded that they do not bind to DNA.

III. Inhibitors of RNA Synthesis Interacting with RNA Polymerases

In this chapter the antibiotics are subdivided into those interacting with RNA polymerase from prokaryotic cells, and those interacting with RNA polymerases from eukaryotic organisms. The short introductory survey on prokaryotic and eukaryotic RNA polymerases is based on the reviews of BURGESS (1971), BLATTI et al. (1970) and KEDINGER et al. (1971). The protomeric form of RNA polymerase, which was first isolated from *E. coli*, contains the following polypeptide chain subunits: one β' subunit; one β subunit; two α subunits; and one σ subunit. This RNA polymerase is designated as complete or holoenzyme ($\alpha_2\beta\beta'\sigma$), which can be separated into two functional parts; a core enzyme ($\alpha_2\beta\beta'$) which is able to synthesize RNA but lacks the ability to initiate RNA synthesis at specific points along the DNA chain and the sigma factor which acts catalytically and allows the initiation of RNA chains at specific sites.

In addition to the sigma factor, other factors which do not bind tightly to the core enzyme have been found to be involved in the transcription process. The ψ-factor is apparently responsible for rRNA synthesis (TRAVERS et al., 1970); the catabolite gene activator protein (CAP) in combination with cyclic AMP, probably permits the transcription of catabolite-sensitive genes (ZUBAY et al., 1970). The ρ-factor seems to be involved in the termination of RNA chains (ROBERTS, 1969).

The subunit composition of *E. coli* RNA polymerase makes it one of the most complex enzymes yet studied. Work on the other RNA polymerases suggests that this complexity is not restricted to the *E. coli* enzyme. Gel patterns from *Azotobacter vinelandii* appear identical to those of *E. coli*, while those of *B. subtilis* reveal the same number and types of subunits. Although some subunits differ in size from their *E. coli* counterparts, preliminary gel patterns of RNA polymerase from higher organisms also show some similarities to the bacterial enzyme.

In contrast to bacteria, higher organisms have been found to contain more than one RNA polymerase these have been designated by BLATTI et al. (1970) as RNA polymerases I, II, III and IV. RNA polymerase I is located in the nucleoli, activated by Mg^{2+} and produces an RNA of ribosomal type. Its major product is most likely 45S-ribosome precursor RNA. RNA polymerase II is localized chiefly in the nucleoplasm, is activated by Mn^{2+} and ammonium sulfate, and produces DNA-like RNA (WIDNELL and TATA, 1966; POGO et al., 1967). The role of polymerase III remains to be clarified; possibly it synthesizes tRNA. RNA polymerase IV may be a unique mitochondrial enzyme. Up to now five multiple DNA-dependent RNA polymerases have been detected in the nuclei of eukaryotic cells. According to the nomenclature of KEDINGER et al. (1971), these are A I, A II, A III, B I and B II, the A forms being insensitive, and the B forms sensitive, to inhibition by α-amanitin. Forms A I and A II are formed in the nucleolus, while the others are located in the nucleoplasm (ROEDER and RUTTER, 1970).

The antibiotics described in this chapter are extremely useful for the discrimination of these different forms of eukaryotic RNA poly-merases. Moreover, as has been shown for rifamycins, they serve as useful tools for the differentiation of specific from unspecific RNA polymerase-binding sites at the DNA template.

A. Rifamycins

1. Origin, Biological and Chemical Properties

Rifamycins are a group of closely related antibiotics produced by *Streptomyces mediterranei*. SENSI et al. (1959) isolated the antibiotics and showed their preferential activity against prokaryotes, particu-larly gram-positive bacteria and *Mycobacterium tubercolosis*. The "rifa-mycin complex" contained five different antibiotics, from which rifa-mycin B - purified in almost a pure state by MARGALITH and PAGANI (1961) - proved to be the most interesting natural rifamycin.

In aerated aqueous solutions, rifamycin B undergoes activation as a result of the formation of a derivative called rifamycin S, which is much more active than rifamycin B. During the activation process rifa-mycin B is oxidized to rifamycin O, which is in turn hydrolyzed to rifamycin S. Rifamycin S can easily be reduced to rifamycin SV (Fig. 55).

The chemical structures of these four rifamycins were elucidated by PRELOG (1963 a, b), OPPOLZER et al. (1964) and BRUFANI et al. (1964). The rifamycins together with streptovaricins and polypomycin form a new unusual class of antibiotics: All of them are ansa compounds con-taining an aromatic ring system which is spanned by an aliphatic bridge. For this reason the name "ansamycins" has been proposed for these antibiotics. Rifamycins can be described as derivatives of an α-aminonaphthohydroquinone. The aliphatic bridge from position 2 to position 5 is composed of acetic acid and propionic acid residues. In position 4 rifamycin B contains a glycolic acid residue. The hydro-gen in position 3 can very easily be substituted by other residues. Thus a synthetic product was derived by substituting this hydrogen by a 4-(methyl)-piperazinyl-iminomethyl residue.

Extensive reviews on the chemical, biological and pharmacological pro-perties of rifamycins have been published by FRONTALI and TECCE (1967), SILVESTRI (1970), GOLDBERG and FRIEDMAN (1971), WEHRLI and STAEHELIN (1971), BINDA et al., (1971) and RIVA and SILVESTRI (1972). There-fore, only a brief survey of the most important facts of the mecha-nism of rifamycin action is presented here.

2. On the Molecular Mechanism of Action

Rifampicin (Fig. 56), the semisynthetic, highly-active derivative of rifamycin B, preferentially inhibits the synthesis of cellular RNA in susceptible bacteria, DNA synthesis remaining unaffected (GOLDBERG and FRIEDMAN, 1971). The rifamycins specifically affect the bacterial DNA-dependent RNA polymerase (E.C. 2.7.7.6) (HARTMANN et al., 1967; UMEZAWA et al., 1968; WEHRLI et al., 1968 a, b; WILHELM et al., 1968).

109

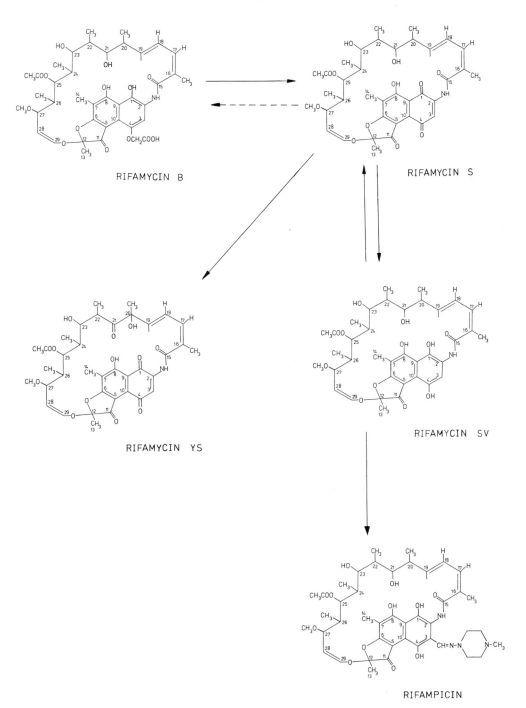

Fig. 55. Structural formulas of various rifamycin derivatives. Symbols: →, chemical reactions; --→, biosynthetic pathways. (According to WEHRLI and STAEHELIN, 1971. From: Bact. Rev. 35, 295 (1971))

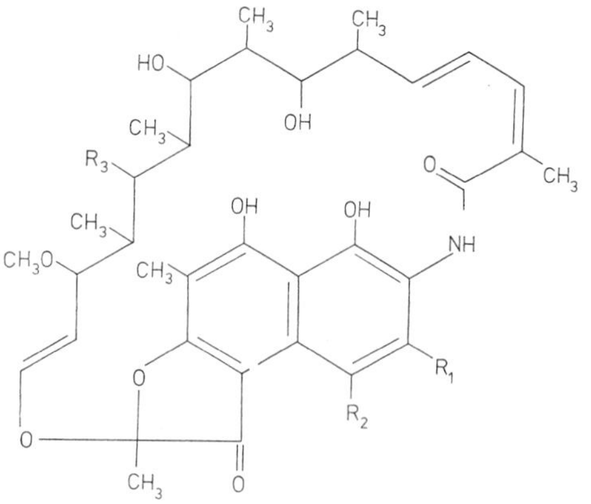

Number	Derivative	Class	R_1	R_2	R_3
--	Rifampicin	A	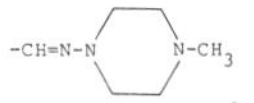 $-CH=N-N\underset{}{\bigcirc}N-CH_3$	$-OH$	$-OOCCH_3$
2	M/14 DA	A	$-H$	$-OCH_2CON\begin{smallmatrix}C_2H_5\\C_2H_5\end{smallmatrix}$	$-OH$
3	M/14	A	$-H$	$-OCH_2CON\begin{smallmatrix}C_2H_5\\C_2H_5\end{smallmatrix}$	$-OOCCH_3$
--	Rifamycin SV	B	$-H$	$-OH$	$-OOCCH_3$
1	AF/AP (N-demethylrifampicin)	B	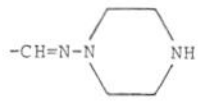 $-CH=N-N\underset{}{\bigcirc}NH$	$-OH$	$-OOCCH_3$
68	AF/013	C	$-CH=NO(CH_2)_3CH_3$	$-OH$	$-OOCCH_3$
97	AF/DNFI	C	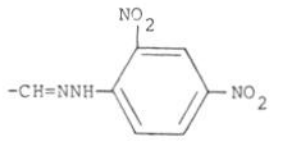 $-CH=NNH-\underset{}{\bigcirc}\begin{smallmatrix}NO_2\\NO_2\end{smallmatrix}$	$-OH$	$-OOCCH_3$
120	AF/ABDP	C	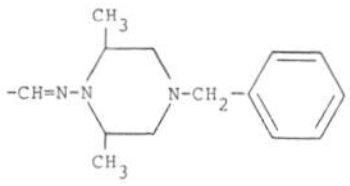 $-CH=N-N\underset{CH_3}{\overset{CH_3}{\bigcirc}}N-CH_2-\bigcirc$	$-OH$	$-OOCCH_3$

Fig. 56. Rifamycin derivatives. Numbers in the first column represent our indentifying symbols, while the symbols in the second column represent the manufacturer's name or code. Class A derivatives are ineffective inhibitors of human DNA polymerases and of RNA tumor virus reverse transcriptase and of infectivity of these viruses. Class B derivatives inhibit these enzymes with moderate potency. Class C compounds are potent inhibitors of these enzymes and of viral infectivity. (According to SMITH et al., 1972. From: Nature New Biol. 236, 167 (1972))

In the cell-free system, inhibition of the RNA polymerase reaction can be overcome by increasing the amount of enzyme, but not by increasing the amounts of the other components in the reaction system, indicating that the antibiotic interacts with the enzyme.

a) Interaction with RNA Polymerase *in Vitro*

Rifampicin forms relatively stable complexes with RNA polymerase (WEHRLI and STAEHELIN, 1971). The complex between RNA polymerase and

rifampicin can be isolated by mixing the free enzyme with the antibiotic and subsequently passing the mixture through a Sephadex column
(WEHRLI et al., 1968 a, b). Quantitative measurements by WEHRLI and
STAEHELIN (197o) reveal that both the holoenzyme and the core enzyme
bind 1 mole of rifampicin per 1 mole enzyme. Thus the σ-factor is not
involved in complex formation - as was postulated by Di MAURO et al.
(1969) and TRAVERS and BURGESS (1969).

Mutants of *E. coli* have been found to be totally resistant to the
action of the drug. Concentrations up to 1,ooo times higher than those
inhibiting a sensitive RNA polymerase showed no effect on the resistant enzyme. When the enzyme from a resistant *E. coli* strain was separated by electrophoresis, the β-subunit was found to have an altered
electrophoretic mobility (RABUSSAY and ZILLIG, 1969). Reconstitution
of the insensitive β-subunit from resistant strains with α and β'
subunits of the sensitive enzyme resulted in a core enzyme which did
not interact with the antibiotic. These results indicate that the β-
subunit is inactivated by the interaction with rifamycins. It has not
been possible to observe direct binding of rifamycin to one of the
purified subunits since, for the separation of the subunits, it has
been necessary to treat the enzyme with 6 *M* urea. Under these conditions the antibiotic is removed and not bound again after removal of
the urea (ZILLIG et al., 197o a, b).

HARFORD and SUEOKA (197o) studied three rifampicin-resistant mutants
of *B. subtilis*. The results suggest that the mutants possess an altered resistant RNA polymerase, and that the location of a probable structural gene for this enzyme is close to the resistance markers for the
3o S ribosome subunit.

To determine which part of the rifamycin molecule is responsible for
the specific action on the RNA polymerase of *E. coli*, the influence
of various derivatives of rifamycin SV has been compared (WEHRLI and
STAEHELIN, 1969). Changes in the aromatic part of the molecule (see
rifamycin SV and rifampicin) have little effect whereas modifications
of the aliphatic bridge influence the inhibitory potencies. Changes
leading to an altered stereochemistry of the ansa ring may impair the
fit of the molecule to the acceptor site on the enzyme. Strikingly,
rifamycin YS, whose structure is identical with that of rifamycin S
with the exception of a keto instead of a hydroxyl group in position
21 and an additional hydroxyl group in position 2o, is completely unable to bind the enzyme. It is concluded that the correct shape of
the ansa ring and the substituents in position 2o and 21 seem to be
most important for the potency of binding to the enzyme. Since the
stability of the RNA polymerase-rifampicin complex is not absolute
and the bound antibiotic slowly exchanges with the free antibiotic,
rifamycins cannot be bound covalently to the enzyme. Therefore, hydrogen bonds, π-π bond interactions between the naphthoquinone ring and
aromatic amino acids of the enzymes, are assumed to be involved in
the formation of the complex.

The capacity of RNA polymerase to bind rifampicin depends on the conformational integrity of the enzyme (LILL and HARTMANN, 1972, 1973).
Conditions which lead to conformational changes e.g. freezing and
thawing in 1 M LiCl, treatment with guanidinium chloride destroys the
binding affinity for the inhibitor. From these results the investigators expected that rifampicin does not bind to the isolated subunits
in solution with a strength comparable to that of native enzyme. Upon
gel electrophoresis of a mixture of isolated RNA polymerase subunits
and [3H]rifampicin no radioactive rifampicin was found to be bound to
the subunits, whereas native or reconstituted RNA polymerase binds

rifampicin stoichiometrically under the same conditions. LILL and HARTMANN suggest, that if the antibiotic is bound to only one subunit in the native RNA polymerase, this subunit may acquire its proper conformation for binding only in combination with the other subunits in the native enzyme. That several subunits may form the binding site seemed less likely with regard to the fact, that rifampicin resistant RNA polymerase carry the resistance always in the β subunit. RNA polymerase contains a large number of SH-groups which are not directly involved in the binding of rifampicin.

b) On the Mechanism of Inhibition of RNA Polymerase Reaction

The RNA polymerase reaction may be roughly divided into four phases: 1. the binding of the enzyme to DNA; 2. the initiation phase; 3. the elongation phase and 4. the termination phase. Studies on the individual steps of enzymic RNA synthesis suggest that once chain initiation has been accomplished, RNA synthesis becomes resistant to rifampicin (SIPPEL and HARTMANN, 1968, 1970).

Rifampicin does not prevent the binding of RNA polymerase to DNA UMEZAWA et al., 1968; SENTENAC and FROMAGEOT, 1969). WU and GOLD-THWAIT's (1969 a, b) results indicate that rifampicin inhibits the binding of purine ribonucleoside triphosphate to the enzyme at a site which is probably responsible for the binding of the initial purine nucleoside triphosphate, which forms the 5'-terminus of the nascent RNA chain. SIPPEL and HARTMANN (1970) postulated that rifampicin only binds to the free enzyme or to a transient enzyme-DNA complex before a conformational change to an activated enzyme-DNA complex has taken place, which causes tight binding of the enzyme to the DNA template and allows RNA chains to be initiated.

The investigations on the dependence on rifampicin binding to RNA polymerase on the conformation of the enzyme lead LILL and HARTMANN (1973) to assume that during chain elongation the conformation of RNA polymerase is altered in such a way that it is no more accessible for the antibiotic.

HINKLE and CHAMBERLIN (1970) proposed a model of the interaction of *E. coli* RNA polymerase with T_7 DNA in which a highly stable complex between the polymerase (holoenzyme) and T_7 DNA is formed. In this model the sequence of steps in site selection by RNA polymerase involves 1. reversible weak binding to random sites on DNA, leading ultimately to 2. attachment to a site at or near the promoter site on the T_7 molecule and 3. a reaction to form a highly stable complex. Rifampicin may act to block one of these steps in the formation of the specific complex between holoenzyme and the T_7 promoter site. BAUTZ and BAUTZ (1970 a, b) studied the effect of rifampicin on transcription with T_7 DNA and suggested that only the "enzyme promoter" complex was resistant to rifampicin. SIPPEL and HARTMANN (1970) found that both RNA polymerase holoenzyme and core enzyme form resistant complexes although the amount of complex resistant to rifampicin was lower when the core polymerase was used. Complexes formed between core polymerase and T_2 DNA or poly [d(A-T)] are far more resistant to rifampicin than is the T_7 DNA-holo enzyme complex.

According to HINKLE et al. (1972) rifampicin can attack both the complex of core enzyme and the complex of holoenzyme with T_7 DNA. The relative higher resistance of RNA polymerase holoenzyme-T_7 DNA complexes to attack by rifampicin - when the drug is added together with the nucleoside triphosphates - is explained with a rapid rate of chain initiation by this complex. During chain initiation the rifampicin-sensitive binary enzyme-T_7 DNA complex is converted to a ternary complex which is completely resistant to rifampicin.

STRAAT and TS'O (197o) studied the inhibitory action of rifampicin on DNA-dependent RNA synthesis and on poly U-dependent poly A synthesis in the *Micrococcus luteus* RNA polymerase system. The DNA-dependent RNA polymerase reaction is initiated with purine nucleoside triphosphates, the RNA polymerase system with homopolynucleotide templates is initiated by the introduction of oligonucleotides complementary to the template. Nucleoside triphosphates protect DNA-directed RNA polymerase from inhibition by rifampicin. Oligomers in the poly U-directed synthesis of poly A did not exhibit similar protective activity. The sensitivity of the initiation reaction to rifampicin is not diminished by the presence of $A(pA)_4$. The role of the oligomers is presumably that of initiators; thus initiation complexes between enzyme, template and oligomers may be formed. Addition of rifampicin before the oligomer prevents the utilization of the oligomer as an initiator - probably by preventing the formation of the initiation complex. However, the addition of rifampicin after the addition of the oligomer and the presumed formation of the initiation complex still prevent the utilization of oligo-A as an initiator. Thus, either the oligomer does not compete at the same site as rifampicin and nucleoside triphosphates or it does not elicit the conformational change that renders the enzyme insensitive to rifampicin. Alternatively, perhaps the oligomer is displaced by rifampicin. It is concluded that oligomers are not capable of bypassing the initiation phase of poly U-directed AMP incorporation in the presence of rifampicin.

Experimental results from LILL and HARTMANN (1973) clearly show that during chain elongation very little rifampicin is bound to the enzyme. Simultaneously enzymatic activity is not affected by the inhibitor. RNA can not protect the enzyme against the attachment of rifampicin although the binding capacity is reduced. Treatment of the complex of enzyme and antibiotic with RNA does not release rifampicin from the complex. DNA alone in the complete absence of substrate can protect RNA polymerase against inhibition by rifampicin (LILL et al., 197o; KERRICH-SANTO and HARTMANN, 1972; BAUTZ and BAUTZ, 197o a). A strict experimental discrimination between chain initiation (enzyme and template in the presence of just one kind of nucleoside triphosphate prior to the first phosphodiester bound) and limited chain elongation is still difficult.

c) Effect on RNA in Intact Cells or Organelles

Prokaryotes: The mechanism of inhibition of bacterial growth by the rifamycins has been widely studied. The findings thus far obtained indicate that the major mode of action of these antibiotics is the interaction with the DNA-dependent RNA polymerase, thus preventing direct DNA-dependent synthesis of RNA within the cells. For summary see WEHRLI and STAEHELIN (1971). For several bacterial strains with different sensitivities to rifampicins it has been shown that the sensitivity was strictly correlated to the abilitiy of the RNA polymerase to bind the antibiotic. Despite the high specificity of rifampicin for the DNA-dependent RNA polymerase reaction within the cells, studies on the growth-inhibitory effect of these antibiotics do not allow the conclusion that resistant strains always contain resistant RNA polymerases. ROMERO et al. (1971) showed that the over-all level of resistance to rifampicin in *E. coli* is controlled by two parameters: 1. the intrinsic sensitivity of RNA polymerase and 2. the permeability of the cell.

The kinetics of RNA and protein synthesis in normal and EDTA-treated
E. coli were measured by REID and SPEYER (197o). With the intact cells,
2oo-fold higher external concentrations of rifampicin were needed to
produce a level of inhibition to that observed in EDTA-treated cells.
This indicates a permeability barrier to rifampicin in *E. coli* and
not the occurrence of a rather rifampicin-resistant RNA polymerase.

Among the many synthetically prepared rifamycin derivatives there
exists a group of compounds derived from the parent substance rifamycin SV by substitution with cyclic secondary amines in position 3
(KNÜSEL et al., 1969). The antibacterial properties differ appreciably from those of rifampicin: firstly, the rate at which resistant
mutants occur among gram-positive organisms is at least a hundred
times lower than it is with rifampicin, although these derivatives
are also highly active against sensitive strains. Secondly, such derivatives display weaker activity than rifampicin against gram-negative bacteria. Thirdly, they are capable of inhibiting rifampicin-
resistant mutants of *Staph. aureus* in concentrations of 1-1o mg/ml. In
concentrations of up to 2oo µg/ml, derivatives from this group have
no influence on the polymerase from resistant cells of *Staph. aureus in
vitro*.

In the course of a large screening program of synthesis and search of
new rifamycin derivatives at Lepetit S.p.A Milano many new rifamycin
derivatives were obtained. A rifamycin derivative (AF/O13, O-n-octyl-
oxime of 3-formyl rifamycin SV) was found to inhibit *in vitro* RNA polymerase from a rifampicin resistant mutant of *E. coli* (RIVA et al.,
1972 a). At inhibitory concentrations several hundred molecules are
bound to one enzyme molecule. These results are not consistent with a
specific binding site at the RNA polymerase for that derivative. The
specificity of the inhibitory effect of rifamycin derivatives on different nucleic acid polymerases is described by GERARD et al. (1973).

In animal cells rifamycin-derivatives have been found to possess
rather unspecific properties on the metabolism of macromolecules. Experiments with intact HeLa cells showed that the synthesis of RNA, the
processing of rRNA, the synthesis of DNA and the uptake of nucleosides
were inhibited by rifamycins (BUSIELLO et al., 1973). In addition the
same derivatives inhibited polyphenylalanine-synthesis in cell free
systems. It is concluded, that the specificity of effects obtained
at concentrations necessary to interfere with normal or virus directed processes in animal cells can be seriously questioned.

The synthesis of 5S ribosomal RNA in *E. coli* treated with rifampicin
was studied by DOOLITTLE and PACE (197o). In these investigations rifampicin was used as a tool to measure the size of transcriptional
units, in particular for 5S ribosomal RNA. The results suggest that
5S RNA is formed by a transcription unit carrying about 2o times the
genetic information necessary for the production of one molecule.
WINSTEN and HUANG (1972) reported the isolation - from log-phase cells
of *B. subtilis* - of a protein-DNA complex which is able to transcribe
RNA chains and which can be inhibited by rifampicin. They suggest
that the RNA polymerase exists in the complex in a dissociable pre-
initiation state. The synthesis observed represents new rounds of
transcription of endogenous DNA initiated *in vitro*. The RNA trans-
scribed by the complex was found to be greatly enriched in ribosomal
RNA sequences.

Eukaryotes: In contrast to bacterial systems, in which only a single
species of RNA polymerase is known, eukaryotic cells contain a varie-

ty of different molecular forms. (For summary see BLATTI et al. (197o).) The eukaryotic RNA polymerases have been designated as RNA polymerases I, II, III and IV. There is increasing evidence that the different species of eukaryotic RNA polymerases occur in different locations in the cells. RNA polymerase I is found in the nucleolus and RNA polymerases II and III are located primarily in the nucleoplasm. RNA polymerase IV may be an unique mitochondrial enzyme.

It is now widely accepted that the rifamycins are inhibitors of RNA polymerase from mitochondria and chloroplasts, but are not active against other RNA polymerases from eukaryotic cells (HARTMANN et al., 1967; WEHRLI et al., 1968 a, b; UMEZAWA et al., 1968; WINTERSBERGER, 197o; SHMERLING, 1969; SURZYCKI, 1969). The lack of inhibition observed when RNA synthesis is studied in the presence of intact mitochondria indicates that the mitochondrial membrane is impermeable to the antibiotic. This suggests that the mitochondrial polymerase *in vivo* may not be affected by rifampicin (GADALETA et al., 197o).

GRANT and POULTER (1973) isolated mitochondria from the myxomycete *Physarum polycephalum*. RNA-synthesis in mitochondria was almost completely inhibited by rifampicin at low levels. The product of rifampicin inhibitable transcription in mitochondria was isolated and is suggested to have mRNA function. In this case rifampicin was used to prove that mitochondrial DNA is transcribed and codes for specific proteins.

The effect of rifamycin antibiotics on algae has been studied by RODRIGUEZ-LÓPEZ et al. (197o). The results reported in this work support the suggestion that in blue-green algae RNA polymerase is the main target of the rifamycin antibiotics.

3. Particular Activities in Biological Systems

a) Bacteriophages, Viruses and Episomes

In any case in which the transcription of a host-specific RNA by a rifampicin-sensitive bacterial RNA polymerase - and thus the synthesis of a host specific protein - is involved in the production of mature DNA or RNA phages, rifampicin will inhibit the growth or maturation of the bacteriophage. Rifampicin was shown to decrease considerably the yield of DNA or RNA phages in bacteria containing a rifampicin-sensitive RNA polymerase, and failed to exhibit an inhibitory effect on the growth of phages in bacteria containing a resistant polymerase. (For summaries see WEHRLI and STAEHELIN (1971) and RIVA and SIVLESTRI (1972)).

MEIER and HOFSCHNEIDER (1972) reported that the replication of RNA phage M_{12} proceeds in a normal way only some 2o min after addition of the drug; later on both phage-specific RNA and protein synthesis simultaneously come to a halt. The authors suggest that rifampicin causes a depletion of one or more host-controlled factors necessary for both phage-RNA and protein synthesis. This would then agree with the observation that rifampicin does not affect M_{12}-dependent protein synthesis in a cell-free extract of *E. coli* (KONINGS, unpublished results). The formation of M_{12} in an *E. coli* HFr strain bearing a rifampicin-resistant RNA polymerase is also unaffected by rifampicin (STAUDENBAUER and HOFSCHNEIDER, 1972). ENGELBERG (1972) studied the effect of rifampicin on the two steps of MS_{12} replication *in vivo*: RNA minus strands are synthesized on the parental template at a reduced rate, while the synthesis of progeny plus strains is completely inhibited. The effect of the drug is not immediate, but a period of approximately 15 min is required for the drug to exert a maximal inhibitory effect on both stages of RNA replication. An interesting aspect is discussed: the inhibition of RNA phage replication by rifampicin *in vivo* may reflect the competition of RNA phage replicase and DNA-dependent RNA polymerase for a protein(s) similar to the Ψ_r factor, which might have a role in the over-all control of the bacterial RNA transcription process. The results presented by ENGELBERG (1972) indicate that a host-specified factor(s) (or subunit(s) of RNA replicase) is(are) only essential for the synthesis of progeny plus strands and is(are) not absolutely required for the synthesis of minus strands on the parental plus-strand template.

In contrast to bacteria and their viruses, in which the action of rifamycins can be reasonably well understood, the effects on mammalian viruses are very heterogeneous, and no general picture of the mechanism of action has yet emerged. The action of the antibiotic on mammalian viruses is, however, quite different from that on bacteria. Two differences among many have been observed: a) certain specific derivatives are markedly active and b) very high concentrations of the antibiotic are required to obtain any effect. Since very little can be said about the mechanism by which rifampicin inhibits the growth of some DNA or RNA viruses, the results should be briefly summarized: rifampicin selectively inhibits vaccinia virus (HELLER et al., 1969). This DNA virus has associated with it DNA-dependent RNA polymerase activity (KATES and McAUSLAN, 1967). More detailed studies of MOSS et al. (1969) showed that the complete viral genome and a large number of viral proteins were made in rifampicin-treated cells. Yet the drug prevents assembly of the virus as was confirmed by electron microscopy. The authors conclude from their biochemical and electron microscopic studies that rifampicin interferes with the assembly of DNA and proteins into virus particles at a stage when the virus envelope is formed. It is speculated that the primary action of rifampicin is to block virus assembly so that virus-specific RNA polymerase is not incorporated into the particles.

GALLO et al. (197o) and GALLO et al. (1971) reported that an RNA-dependent DNA polymerase in fresh human leukemic cells was inhibited by N-demethylrifampicin, but not by rifampicin. The enzyme resembles viral reverse transcriptase in its sensitivity to these two inhibitors (GURGO et al., 1971; GALLO et al., 1971).

Derivatives of rifampicin were also found to inhibit the replication of certain DNA viruses (SUBAK-SHARPE et al., 1969), but apparently by a mechanism different from that in bacterial systems (SIPPEL and HARTMANN, 1968; MOSS et al., 1969). A few derivatives of rifamycin SV were at first shown to inhibit DNA polymerases (reverse transcriptases) of oncogenic viruses (GURGO et al., 1971). Later it became clear that the rifamycin derivatives could be classified into groups with widely varying effects on reverse transcriptase (YANG et al., 1972; GURGO and GREEN, 1972; SMITH et al., 1972). These investigators studied more than 2oo rifamycin derivatives and classified them into 3 groups designated A, B and C (Fig. 56). The class A derivatives are ineffective for both viral reverse transcriptase and purified human DNA polymerase, class B derivatives inhibit the enzymes with moderate potencies, while class C derivatives were found to be potent inhibitors. Several inhibitors from classes B and C efficiently block murine sarcoma virus-induced transformation of rat cells under non-cytotoxic conditions, while several class A derivatives do not inhibit transformation (TING et al., 1972).

With the chemical modifications of rifampicin a new and promising era
has begun which, hopefully, will provide us with selective inhibitors
for DNA polymerases, reverse transcriptases and RNA polymerases.

Circular "extrachromosomal DNA molecules" such as F, F' and R factors
code for a number of functions, e.g. pilus formation, fertility, autonomous replication, restrictions of certain phages and exclusion of
other extrachromosomal elements. Very low concentrations of rifampicin eliminate staphylococcal plasmids (JOHNSTON an RICHMOND, 197o).
An inhibitory effect was also observed on the transfer of R factors
in enterobacteria (KRCMÉRY and JANOUSKOVA, 1971). RIVA et al. (1972 b)
reported the inhibitory effect of rifampicin on the expression of episomal genes in *E. coli*; the mechanism of this effect is as yet unknown.

CLEWELL et al. (1972) reported on the direct inhibitory effect of
rifampicin on Col E_1 plasmid DNA replication in *E. coli*. In later experiments CLEWELL and EVENCHICK (1973) represent a further characterization of this phenomenon. Studies involving actively growing cells
demonstrated that rifampicin effectively inhibits the incorporation
of radioactive thymine into Col E_1 DNA while allowing chromosomal DNA
to complete the current round of replication. Since we know that replication starts with short peaces of ribonucleotides on the DNA template, the primary effect of rifampicin may be an inhibition of the
start of DNA replication with short RNA pieces.

b) Trachoma Agent

The infectious elementary bodies of trachoma agent belong to the chla-
mydiaceae, which are considered unusually small bacterial cells. They
contain DNA, RNA and ribosomal subunits and grow as parasites in mammalian cells (SAROV and BECKER, 1968). Rifampicin inhibits the replication of trachoma agent and the related psittacosis-lymphogranu-
loma agents in cell cultures *in vitro* and in embryonated eggs *in vivo*.
The hydrazone side chain present in rifampicin is necessary for its
antitrachoma and antiviral acitivity. Since the concentration which inhibits replication of trachoma agent is also much higher than that necessary for inhibition of bacterial RNA polymerase, it is possible that,
at high doses, rifampicin acts by inhibiting selectively the synthesis
of proteins specific to trachoma agent.

B. Streptovaricin

1. Origin, Biological and Chemical Properties

Streptovaricin was isolated by SIMINOFF et al. (1957) from *Strepto-*
myces spectabilis and consisted of a mixture of at least five compounds
(WHITEFIELD et al., 1957; SOKOLSKI et al., 1958; MIZUNO et al., 1968).
This antibiotic complex is active mainly against gram-positive bac-
teria (SIMINOFF et al., 1957) including *Mycobacterium tuberculosis* (RHU-
LAND et al., 1957).

Like the rifamycins, the streptovaricins are ansa compounds containing a two-membered ring system with quinone properties. The streptovaricin complex consists of a mixture of certain streptovaricins

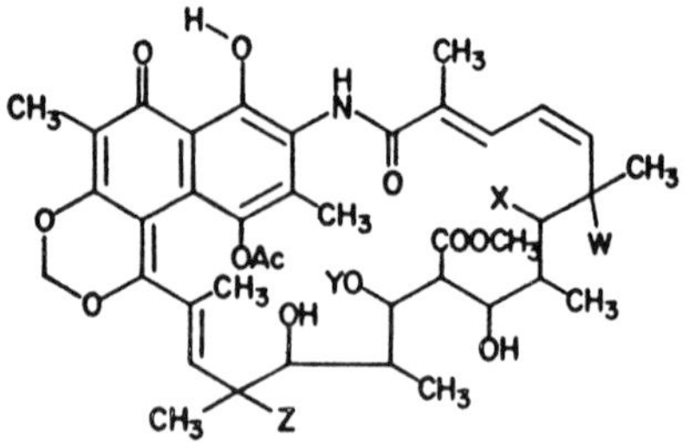

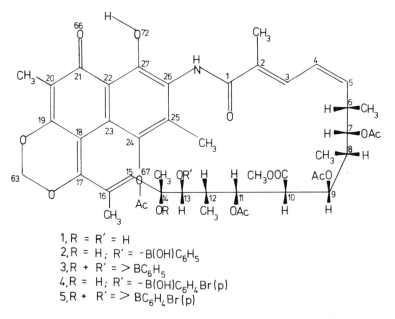

1(A), W = OH; X = H, OH; Y = Ac; Z = OH
2(B), W = H; X = H,OH; Y = Ac; Z = OH
3(C), W = H; X = H,OH; Y = H; Z = OH
4(D), W = H; X = H,OH; Y = H; Z = H
5(E), W = H; X = =O; Y = H; Z = OH
7(G), W = OH; X = H,OH; Y = H; Z = OH

Fig. 57. Structures of streptovaricins. (According to RINEHART et al., 1971. From: J. Amer. Chem. Soc. 93, 6273 (1971))

1, R = R′ = H
2, R = H; R′ = -B(OH)C₆H₅
3, R + R′ = > BC₆H₅
4, R = H; R′ = -B(OH)C₆H₄Br(p)
5, R + R′ = > BC₆H₄Br(p)

Fig. 58. Structures of streptovaricin C triacetate and derivatives used for X-ray crystallographic determination. (According to WANG et al. From: J. Amer. Chem. Soc. 93, 6275 (1971))

(assigned the letters A through G). The structures of the strepto-varicins (Fig. 57) were elucidated by RINEHART et al. (1971). The full structure was solved by X-ray crystallography with the heavy atom method of WANG et al. (1971). Of particular interest is the observa-tion.that the relative configuration at every comparable chiral cen-ter of compound 5 (Fig. 58) is identical with that in rifamycins B and Y and tolypomycin, although the geometry of the dienamide unit is reversed (trans,cis for compounds 1-5 versus cis, trans for rifamycins B and Y (WANG et al., 1971)). Slightly revised structures for streptovarone and streptovaricin C (compounds 3 and 5) have recently been presented by SASAKI et al. (1972). Closely related to the rifamycins and streptovaricins are tolypomycins (KAMIYA et al., 1969; KISHI et al., 1969 a, b) and geldanamycins (DeBOER et al., 197o).

2. On the Molecular Mechanism of Action

Kinetic studies of the inhibition by streptovaricin of the RNA polymerase reaction indicated that the inhibition is non-competitive with DNA and CTP. This indicates that the action of streptovaricin is neither to bind to DNA primer, nor to compete for the same sites of the enzyme to which the primer DNA or CTP attaches. As in the case with rifamycins the streptovaricins do not inhibit the DNA-dependent RNA polymerase reaction once the complex between DNA and enzyme is formed. The tolypomycins and geldanamycins competitively inhibit the binding of rifamycins to RNA polymerase, thus indicating that the binding site is the same for all three groups of antibiotics (WEHRLI and STAEHELIN, 1972). In general, streptovaricins and rifamycins show closely related modes of action. An RNA polymerase from *E. coli* which was resistant to rifampicin was also resistant to streptovaricin (NITTA et al., 1968). Resistance to streptovaricin has been shown to be due to an altered RNA polymerase (YURA and IGARASHI, 1968; YURA et al., 197o).

Streptovaricin, like rifampicin, does not inhibit the RNA polymerase from higher organisms (MIZUNO et al., 1968). Therefore streptovaricins exhibit low toxicity for animals.

MIZUNO and NITTA (1969) reported that phage T4 RNA synthesis in streptovaricin-sensitive *E. coli* is inhibited by streptovaricin at any time after infection, but that in streptovaricin-resistant cells, phage RNA synthesis is not inhibited at all. The following conclusions were drawn from the experiments: a) during the early and the late periods of phage growth, *E. coli* RNA polymerase functions in phage RNA synthesis; and b) if *E. coli* RNA polymerase is modified by the phage infection - as was suggested by polyacrylamide gel electrophoresis - the structural characteristics of the host enzyme in relation to streptovaricin sensitivity is conserved. Similar observations were made with rifampicins (see appropriate section in this book). TAN and McAUSLAN (1971) showed that streptovaricin D depresses uridine transport into cellular RNA and viral mRNA. This effect is not restricted to bacteria but was found also to occur in HeLa cells. From the different streptovaricins occurring in the complex, streptovaricins A, B, C, D and F, for comparison, also rifampicin were tested. The experiments reveal that only streptovaricin D inhibits the incorporation of uridine into acid-precipitable material from HeLa cells. The inhibition of nucleoside incorporation into cells by streptovaricin D occurs rapidly. It is competitive and probably affects the active transport of nucleosides into cells - as is suggested by the speed of inhibition. Alternatively, streptovaricin D may inhibit phosphorylation of nucleosides and thus the incorporation into cells. Phosphorylation of nucleosides at the cell surface has been suggested as a mechanism for the uptake of nucleosides (PIATIGORSKY and WHITELEY, 1965). The inhibition is rapid and reversible, suggesting that streptovaricin D is closely, but loosely, associated with the cell membrane.

BROCKMAN et al. (1971) reported that the streptovaricin complex is an extremely potent inhibitor of the reaction by which DNA is transcribed from the RNA template resident in purified murine leukemia viruses. In the system they used, murine leukemia virus grown in MLS-V9 cells, streptovaricin C and rifamycin SV were also quite active, while streptolydigin and rifampicin were relatively poor inhibitors. The experiments were extended by CARTER et al. (1971) and BORDEN et al. (1971). They tested the effect of streptovaricins on the efficiency of transformation by Maloney murine sarcoma-leukemia complex (MSV) *in vitro*, and on the splenomegaly induced by Rauscher leukemia virus in mice. Various streptovaricins and a number of rifamycins were compared. Cell multiplication was not suppressed by 2 µg/ml streptovaricin complex, by 2o µg/ml streptovaricin D or by 8o µg/ml streptovaricin A. But 3-formyl rifamycin SV, was cytotoxic at 1 µg/ml, a concentration at which this compound inhibits reverse transcriptase. Those streptovaricins which are active inhibitors of formation of the MSV-induced foci do so at concentrations which do not impair the diversion rate of normal cells. This suggests that the effect on blockage of transformation is due to a selective inhibition of RNA-dependent DNA polymerase, because inhibition of focus formation parallels the inhibition observed on the enzyme assay. Focus inhibition depends critically on the time of drug addition, which is indicative of inhibition of an early viral-specific function. Moreover, the inhibition of transformation by streptovaricin was enhanced if the cells were exposed simultaneously to interferon, which inhibits the translation of viral messenger RNA, thus suggesting that enhanced inhibition of focus formation can be achieved by simultaneous blockage of transcription and translation. On the other hand, streptovaricins can reduce the cellular uptake of adenosine, uridine and thymidine and in this way interfere with nucleic acid synthesis.

Streptovaricins inhibit tumor induction by Rauscher leukemia virus in mice. Mice were fed streptovaricin, such that the concentration of the drug in the serum was from 1-3 µg/ml. This protected to a significant extent against splenomegaly induced by RLV. Since transformation was inhibited without influencing the division of rapidly proliferating normal cells, it was assumed that streptovaricins act as inhibitors of tumor virus reverse transcriptase.

C. Streptolydigin

1. Origin, Biological and Chemical Properties

Streptolydigin resembles rifamycins and streptovaricins with respect to its inhibitory effect on bacterial RNA polymerase. Streptolydigin, produced by *Streptomyces lydicus*, was isolated by DE BOER et al. (1955/56), and characterized by EBLE et al. (1955/56). RINEHART et al. (1963) elucidated the structure of streptolydigin (Fig. 59).

Although streptolydigin does not belong to the ansa compounds it contains a considerable part of the ansa ring, including the two oxygen functions at positions 21 and 23 of the rifamycins, and the tetragonal carbon atom 21 which has been shown to be responsible for the activity of rifamycin. Streptolydigin is identical with portamycin.

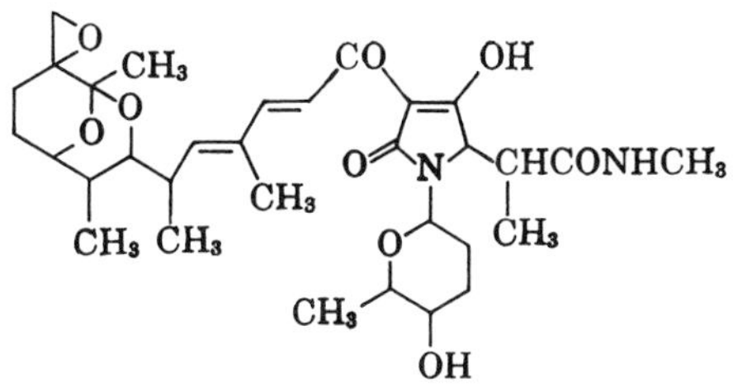

Fig. 59. Structure of streptolydigin. (According to RINEHART et al., 1963. From: J. Amer. Chem. Soc. 85, 4o39 (1963))

2. On the Molecular Mechanism of Action

Like the rifamycins and streptovaricins, streptolydigin specifically interacts with bacterial RNA polymerase, and in analogy to the other inhibitors of RNA polymerase, the β-subunit of bacterial RNA polymerase is involved in binding (ZILLIG et al., 197o). Apart from the marked similarities in the interaction of streptolydigin and rifamycins with RNA polymerase, the mode of inhibition of RNA synthesis is quite different for the two drugs. Rifampicin inhibits initiation of transcription and allows initiated RNA chains to be transcribed. It has been suggested that streptolydigin preferentially inhibits the elongation of RNA chains (SIDDHIKOL et al., 1969; SCHLEIF, 1969).

CASSANI et al. (197o, 1971) studied the effect of streptolydigin on the different phases of cell-free synthesis of RNA-chain initiation, elongation and termination. The concentration of the drug at which inhibition is complete was high (about $10^{-4}M$) compared with that of rifampicin ($10^{-7}M$). In the presence of $1.4 \times 10^{-5}M$ streptolydigin, where the total amount of RNA synthesized was found to be 46% that of the control, the rate of chain growth was reduced to 2.1 nucleotides per second, i.e. 46% of the rate which was observed in the absence of the drug. Thus a direct effect of streptolydigin on the rate of chain growth may account for the inhibitory effect of the drug. The drug is thought to affect the entire population of enzyme molecules to an equal extent. The DNA-enzyme-RNA complex formed during RNA synthesis *in vitro* remains absolutely stable when incubated in the presence of streptolydigin for 5o min at 36°C. The chains of RNA that were growing remain bound to the DNA-enzyme complex. This clearly shows that streptolydigin does not cause a release of RNA chains during synthesis. From the kinetics of the incorporation of nucleoside triphosphates into RNA in the presence of streptolydigin, it is evident that the inhibitor is non-competitive for the binding of ATP and GTP to the RNA polymerase. From the kinetics with UTP it is assumed that the affinity of UTP for RNA polymerase is decreased in the presence of the drug during transcription. On the contrary, the affinity of RNA polymerase for CTP seems to be increased in the presence of the drug.

The addition of $10^{-4}M$ streptolydigin to the polymerase system before incubation completely inhibits the synthesis of TCA-precipitable material. In the presence of the drug RNA polymerase was incapable of promoting the condensation of the first nucleoside triphosphates at the 5'-end of RNA to form a phosphodiester bond. Therefore the drug inhibits the initiation of RNA chains.

From their investigations CASSANI et al. (1971) concluded that streptolydigin binds to the enzyme even in the DNA-RNA polymerase complex.
The effect of the drug is thought to be a tightening of the interaction between enzyme and DNA which results in a decreased ability
of the enzyme to form phosphodiester bonds and to carry on pyrophosphate exchange at the usual rates.

VON DER HELM and KRAKOW (1972),using the d(AT)-directed reaction as
a model system, also showed that streptolydigin stabilizes polymerase-
template interaction. The enzyme-d(AT) complex must be incubated at
37°C for streptolydigin to stabilize the complex. There is a distinct
similarity between the effect of ATP and streptolydigin on the stabilization of the enzyme-d(AT) complex. Mg^{2+} is required for stabilization of the complex by the antibiotic or by ATP. These ligands do
not necessarily bind at the same site on the enzyme, although their
binding sites must be close enough to allow a similar response on
binding either ATP or streptolydigin. VON DER HELM and KRAKOW (1972)
suggest that streptolydigin affects the conformation of the polymerase-
DNA complex at or near the site at which, probably, the 3'-OH of ATP
or at the nascent - pA or pU - during polymerization is held in a
proper juxtaposition to the incoming substrate nucleoside triphosphate
before phosphodiester bond formation.

D. Amanitins

1. Origin, Biological and Chemical Properties

Although the amanitins are not antibiotics, these polypeptides - toxic
components of the poisonous mushroom *Amanita phalloides* (WIELAND and
WIELAND, 1959; WIELAND, 1968) - should be included in this section
on inhibitors of RNA polymerase, because they interact specifically
with RNA polymerase from eukaryotes.

α-amanitin was found neither to inhibit the multiplication of three
species of bacteria, *E. coli*, *Staph. aureus* and *B. subtilis*, nor the replication of two RNA viruses (poliovirus type II and parainfluenza
virus type III) and of two DNA viruses (vaccinia virus and the virus
causing bovine infective rhinotracheitis) (FIUME et al., 1966). A
review of the chemistry and action of amanitins covering the literature through the beginning of 197o has been written by FIUME and
WIELAND (197o).

The structure of α-amanitin was elucidated by WIELAND and GEBERT (1966)
and FAULSTICH et al. (1968) (Fig. 6o). α-amanitin is a cyclic octa-
peptide; the ring contains a sulfoxide bridge originating from a cyst-
eine sulfur attached to the 2-position of the indole nucleus of a
tryptophan unit. The molecule is destroyed by elimination of the sulfoxide bridge or by opening of one of the peptide bonds, which results
in disappearance of activity (WIELAND and WIELAND, 1959).

2. On the Molecular Mechanism of Action

a) Interaction of Amanitins with Eukaryotic RNA Polymerases

In contrast to bacteria, higher organisms contain more than one RNA
polymerase. (For further details see the introduction to Chapter III.)

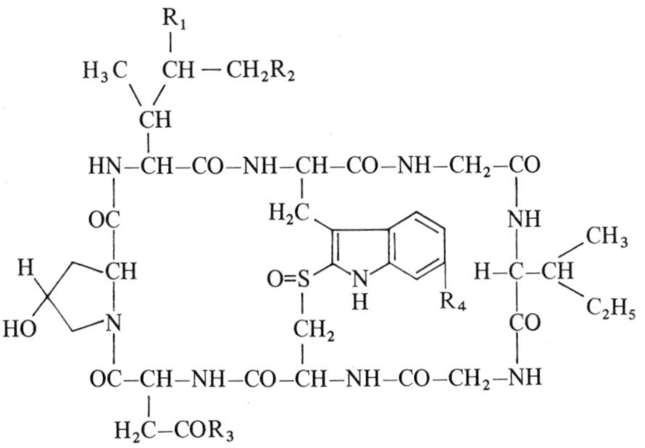

$$
\begin{array}{c}
R_1 \\
| \\
H_3C \quad CH-CH_2R_2 \\
\backslash \diagup \\
CH \\
| \\
HN-CH-CO-NH-CH-CO-NH-CH_2-CO
\end{array}
$$

I

	R_1	R_2	R_3	R_4
a α-Amanitin	OH	OH	NH_2	OH
b β-Amanitin	OH	OH	OH	OH
c γ-Amanitin	OH	H	NH_2	OH
d Amanin	OH	OH	OH	H
e Amanullin	H	H	NH_2	OH

Fig. 6o. Structure of Amanitins. (According to FIUME and WIELAND, 197o. From: FEBS Letters $\underline{8}$, 2 (197o))

NOVELLO et al. (197o) described the effect of α-amanitin on RNA polymerase solubilized from rat liver nuclei. Studies on the inhibition of RNA polymerase by α-amanitin showed that the inhibition was not overcome by increasing the amount of DNA in the reaction medium nor was the apparent K_m for DNA modified in the presence of the toxin. The degree of inhibition decreased as the amount of the enzyme increased. These results suggested that α-amanitin may act directly on the enzyme.

RNA polymerase type I, located in the nucleoli, is activated by Mg^{2+} and produces an RNA of ribosomal type. RNA polymerase II is localized chiefly in the nucleoplasm, is activated by Mn^{2+} and ammonium sulfate and produces DNA-like RNA (WIDNELL and TATA, 1966; POGO et al., 1967).

STIRPE and FIUME (1967) observed that α-amanitin affected RNA polymerase II (which is Mn^{2+} and ammonium-dependent) but had scarcely any effect on the enzyme activity of nuclei assayed at low ionic strengths in the presence of Mg^{2+}. JACOB et al. (197o a,b) also found that in isolated mice liver nuclei, α-amanitin changes the polymerase activity in the presence of Mn^{2+}. On purification of the Mn^{2+}-dependent enzyme, it was found to be highly sensitive against α-amanitin.

More detailed studies revealed different responses of soluble nucleolar RNA polymerase and soluble whole nuclear RNA polymerase to divalent cations and to the inhibition by α-amanitin. From the variable degree of response to amanitin and to Mg^{2+} and Mn^{2+}, it was concluded that the RNA polymerase of nucleoli differs structurally from the chromatin-associated polymerase extractable from whole nuclei. KEDINGER et al. (197o) purified two different types of RNA polymerases from calf thymus and found that α-amanitin was very efficient and

highly specific as an inhibitor of one of the two RNA polymerase activities present in calf thymus. The two types of RNA polymerase activities (A, B) were separated by DEAE cellulose. Polymerase B was significantly inhibited by α-amanitin, while the calf thymus RNA polymerase A remained unaffected.

α-amanitin affects yeast RNA polymerase - which is not inhibited by rifampicin (DEZELEE et al., 1970). The yeast RNA polymerase requires
denatured DNA as a template and Mn^{2+} ions for activity. Although the
enzyme was not absolutely pure, the calculations showed that for complete inhibition, 5oo α-amanitin molecules per molecule of RNA polymerase were necessary - this number is twice the amount required to
inhibit the mammalian enzyme.

LINDELL et al. (1970) were able to separate from sea urchin embryos
three distinct DNA-dependent RNA polymerases with the same location
and function as those described by BLATTI et al. (1970) and ROEDER
and RUTTER (1970). The former suggest that the polymerases have different functions. The effect of α-amanitin on the various nuclear polymerases was tested. α-amanitin specifically inhibits polymerase II
from the nuclei of several organisms, while polymerase I and III activities are not affected. The specific inhibition of polymerase II
by α-amanitin implies a structural difference between this polymerase
and polymerases I and III. Such a difference confirms the conclusion
that polymerases I, II and III are distinct molecules. It is not yet
known whether the three enzymes contain any common subunit.

MEIHLAC et al. (1970) incubated [^{14}C]-methyl-gamma-amanitin with RNA
polymerase from calf thymus. Upon sedimentation through a glycerol
density gradient, the peak of radioactivity coincided with that of
RNA polymerase. The specific binding was thought to give rise to a
stable complex, as no tailing off of radioactivity was evident. A
further separation of the polymerases of calf thymus by polyacrylamide gel electrophoresis (KEDINGER et al., 1971) revealed two α-
amanitin-sensitive RNA polymerases in calf thymus. Enzyme B, described previously, was further purified into two enzymes, B_I and B_{II}. Experiments with [^{14}C]-methyl-gamma-amanitin showed that the label mi-
grated with the two bands of enzymes B_I and B_{II}. Both enzymes were
strongly inhibited by the antibiotic. The physiological functions of
RNA polymerases B_I and B_{II} from calf thymus are unknown. A possible
function for RNA polymerases B_I and B_{II} might be the fact that they
catalyze the synthesis of messenger RNA and nuclear heterogenous RNA.
Very little is known about the mechanism by which α-amanitin interacts with the RNA polymerases and inhibits their function. SEIFART
and SEKERIS (1969) found a dose-response relationship and calculated
that one molecule of enzyme interacts with about 1.5 molecules of inhibitor, making a 1:1 stoichiometry very likely. Preincubation of the
enzyme with α-amanitin allows a large portion of normal chain propagation on addition of CTP. α-amanitin does not selectively act on
the free enzyme but inhibits the enzyme already bound to DNA. SEIFART
and SEKERIS concluded that α-amanitin allows chain initiation, but
almost completely prevents chain propagation. Their conclusion was
confirmed by JACOB et al. (1970 a, b) and LINDELL et al. (1970). The
data obtained so far with a yeast RNA polymerase reveal that α-amanitin inhibits chain propagation, and it is postulated by DEZELEE et al.
(1970) that α-amanitin interferes with the formation of the phosphodiester bond.

b) Effects on RNA Metabolism

The different sensitivity to α-amanitin of the various RNA polymerases in mammalian systems raises the question whether amanitins in-

hibit the synthesis of certain types of RNA within the cell and allow
the production of another type. It has been suggested that the nucleolar enzyme (which is not affected by α-amanitin) catalyzes the synthesis of ribosomal RNA, while the nucleoplasmic enzymes are responsible for the synthesis of DNA-like RNA. JACOB et al. (197o c) examined
the effect of administration of α-amanitin to intact animals on the
synthesis of different species of RNA in rat liver and found, unex-
pectedly, that formation of ribosomal RNA, as well as other species
of RNA, is inhibited by the antibiotic under these conditions. The
difference between the action of the toxin *in vivo* and its effect *in
vitro* is still unexplained. One suggestion is that the synthesis of
ribosomal RNA in the nucleolus may be under the control of an extranucleolar mechanism which is sensitive to α-amanitin.

Isolated nuclei from *Xenopus laevis* tissue culture cells synthesize RNA
when incubated *in vitro*. A major fraction of this RNA is ribosomal RNA
as shown by its specific hybridization to purified ribosomal DNA. The
synthesis of ribosomal RNA is completely insensitive to α-amanitin
(REEDER and ROEDER, 1972). The results indicate, that the RNA-polymerase I which is responsible for rRNA synthesis is resistant to α-
amanitin. RNA polymerase II is sensitive to α-amanitin and is probably
not involved in ribosomal RNA transcription at least in *Xenopus laevis*.
The possible role of RNA polymerase III in rRNA synthesis cannot be
ruled out.

Another possibility is that *in vitro* studies with isolated nuclei,
such as those of NOVELLO et al. (197o), may not reflect the *in vivo*
conditions. It is possible that more purified fractions of RNA polymerase activity which are not sensitive to α-amanitin *in vitro* may be
rendered sensitive by purification procedures.

NIESSING et al. (197o) studied the *in vivo* inhibition of RNA synthesis by α-amanitin. Injection of α-amanitin into rats inhibits both
the synthesis of ribosomal and DNA-like RNA in liver, as measured by
the incorporation of radioactively-labeled orotic acid and orthophosphate into RNA extracted by phenol and subsequently fractionated by
density-gradient centrifugation and MAK chromatography. These experiments and results also indicate that, *in vivo*, α-amanitin has no preferential influence on the synthesis of ribosomal or DNA-like RNA.

The results of TATA et al. (1972) point to a fundamental difference
in the action of α-amanitin on RNA-synthesis in rat liver nuclei *in
vivo* depending on whether RNA is assayed as RNA synthesized *in vivo* or
in vitro. In one series of experiments RNA polymerase activity was assayed in nuclei isolated at different times after the administration
of α-amanitin to the animal. In another series the effect of α-amanitin
in vivo was studied by measuring the overall rate of synthesis *in vivo* of
total nuclear and 4S RNA. The bases of the newly synthesiszed RNA were
analyzed. It was found that α-amanitin *in vivo* has an extremely rapid
but transient effect on one component of RNA polymerase in isolated
nuclei lasting for only about 8o min after its administration to the
animal. The effect of the toxin on RNA synthesis *in vivo* however per-
sists for several hours affecting the formation of ribosomal as well
as non ribosomal RNA.

The effect of α-amanitin on RNA polymerase II, DNA and protein synthesis was studied in regenerating rat liver by MONTECUCCOLI et al.
(1973). The impairment of DNA synthesis which followed the inhibition
of RNA polymerase II is suggested to be a consequence of the blockage
of RNA-synthesis. Thus newly synthesized RNA might be needed as a
primer for DNA synthesis in the living animal as well as in bacteria.

SHAAYA and SEKERIS (1971) studied the inhibitory effect of α-amanitin
on RNA synthesis and on the induction of DOPA-decarboxylase by β-

ecdysone. Six to eight hours after intra-abdominal injection of β-ecdysone into ligated head-pupated calliphora larvae, a significant increase in DOPA-decarboxylase activity was observed. α-amanitin inhibits the induction of DOPA-decarboxylase if given prior to β-ecdysone, but has no effect on the induction process if administered after the hormone. In these experiments α-amanitin was injected three hours before the hormone in order to obtain maximal RNA inhibition by the time the hormone starts to act on RNA synthesis in the epidermis. α-amanitin injected into ligated calliphora larvae largely inhibited the synthesis of heterodisperse RNA, but the synthesis of a small portion of this RNA was unaffected by the drug. The synthesis of ribosomal RNA was completely depressed, but the inhibition of tRNA seemed to begin later, and recovery of synthesis started earlier than that of heterodisperse RNA. The synthesis of tRNA was practically unaffected (SHAAYA and CLEVER, 1972).

The induction of a plant tumor by *Agrobacterium tumefaciens* in the leaves of *Kalanchoe daigremontiana* can be inhibited by rifampicin, which does not influence the synthesis of RNA in the plant cells. The induction of the tumor can also be inhibited by α-amanitin, which does not affect the RNA polymerase of the inducing bacterium. This observation by BEIDERBECK (1972) indicated that during the early phase of transformation the synthesis of a specific plant RNA must occur. The transformation process from a normal cell to a tumor cell in plants thus resembles the virus induced transformation of animal cells.

α-amanitin does not pass the blood-brain barrier, but is highly toxic to rats, when injected intracerebrally, and causes inhibition of RNA polymerase II of isolated brain nuclei. The injection of sublethal doses of α-amanitin impairs the establishment of long-term memory (MONTANARO et al., 1971).

IV. Inhibitors Interferring at the Precursor Level or with Regulatory Processes of Nucleic Acid Synthesis

The close structural relationship of the nucleoside antibiotics to purine and pyrimidine nucleosides and that of amino acid analogs to amino acids have made both groups useful as inhibitors. The most interesting of the nucleoside analogs is cordycepin (3'-deoxyadenosine). This analog of dATP seems to exhibit differential inhibitory activity towards the synthesis of different types of RNA. Formycin, an antibiotic which can be incorporated into RNA instead of adenosine, is also of interest because of certain structural peculiarities. Unlike adenosine formycin can adapt two conformations, syn and anti, which hinders base-pairing between complementary nucleosides in helical nucleic acids.

Scientists all over the world are becoming more and more involved in research on regulatory processes which couple RNA and protein synthesis. FILL et al. (1972) have proposed a model in which the transcription initiation frequency of ribosomal RNA and other stringently regulated genes in bacteria is dependent upon the concentration of a positive control protein, whose transcription-stimulating activity is modulated by the concentration of guanosine tetraphosphate (ppGpp). We present experimental data showing that quinone antibiotics and some synthetic quinones interfere with this regulatory process and can be used to study regulatory mechanisms in DNA transcription.

A. Nucleoside Antibiotics

The nucleoside antibiotics represent a group of compounds structurally related to the purine and pyrimidine nucleotides occurring in nucleic acids. This close structural relationship has made them useful as structural analogs and inhibitors in cell research. A clear, comprehensive review on nucleoside antibiotics was given by SUHADOLNIK (1970). The following brief survey deals with those antibiotics interfering preferentially with nucleic acid synthesis and not with other biochemical pathways in which mono-, di-, or triphosphates of nucleosides are involved.

a) Toyocamycin, Tubercidin, Sangivamycin

These antibiotics are pyrrolopyrimidine nucleasides, which are analogues of adenosine with different substituents at C_5 (Fig. 61). Within the cell the compounds are easily phosphorylated to the respective mono-, di-, and triphosphates. Discovery, production, isolation, physical and chemical properties and structural elucidation are described in detail by SUHADOLNIK (1970). Despite their close chemical relationship the drugs have quite different biological activities.

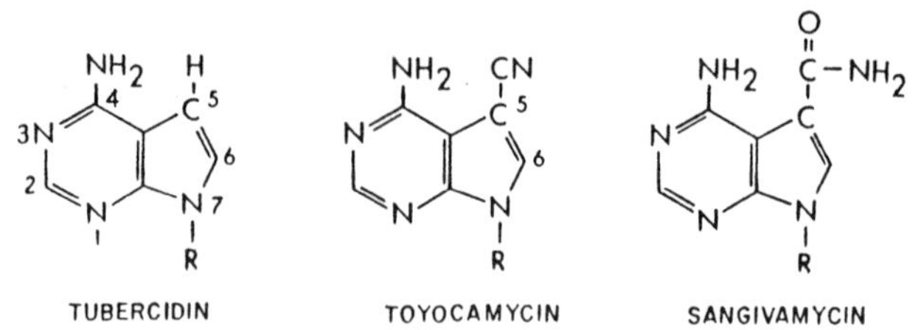

TUBERCIDIN   TOYOCAMYCIN   SANGIVAMYCIN

Fig. 61. The structures of the pyrrolopyrimidine nucleoside antibiotics. (From: SUHADOLNIK, R.J.: Nucleoside Antibiotics, p. 3oo.
New York-London-Sidney-Toronto: J. Wiley and Sons, Inc. 197o)

Toyocamycin was isolated from different strains of *Streptomyces S. toyo-
caensis*: Strains 1922, 1o37, E-212, *S. rimosus* and S. 86 (Toyocamycin
is identical with antibiotic 1o37 and antibiotic E-212, with unamy-
cin B, produced by *S. fungicidus*, with monilin from Streptomyces 6623
and also with vengicides from *S. vendygensis*). Besides its carcinostatic
and cytostatic effects, toyocamycin inhibits *Candida albicans, Penicillum
chrysogenum, Trichophytum interdigitale* and *Mycobacterium phlei*, but has little
effect on most gram-positive and gram-negative bacteria, fungi and
yeast.

In mouse fibroblasts toyocamycin affects the synthesis of RNA, DNA
and protein. Low concentrations of toyocamycin completely inhibit the
formation of 28S and 18S RNA, but have no influence on the synthesis
of 4S and 45S RNA (TAVITIAN et al., 1968). Methyl groups from methio-
nine are incorporated into the 45S RNA, but not into 28S and 18S RNA.
A fourfold increase in the amount of 45S RNA 6 hours after the addition of toyocamycin was found within the cells. This accumulated 45S
RNA could not be converted to 28S and 18S RNA even after the removal
of the nucleoside antibiotic. Thus it is possible that toyocamycin
interferes with the modification and the processing of 45S RNA.

Toyocamycin did not inhibit the synthesis of tRNA and of 5S RNA in mammalian cells in cultures. The 5S RNA was, however, not associated with
the ribosomes (SVERAK et al., 197o). Parallel experiments with cells
from chick embryo revealed that the blockage of rRNA formation and the
concomitant accumulation of 45S RNA precursors only occur in mammalian
cells. In toyocamycin-treated chick embryo cells all stages of rRNA
synthessis were depressed, indicating that the drug inhibits the synthesis of 45S RNA precursor rather than its modification and processing (TAVITIAN et al., 1969).

With partially purified ribonucleotide reductase from *L. leichmannii*
toyocamycin is reduced to the respective 2'-deoxyribonucleotide. This
may explain the occurrence of the compound in DNA of cells treated
with the antibiotic. Since large concentrations of toyocamycin accu-
mulate in Ehrlich ascites tumor cells, the corresponding 2'-deoxy-
ribonucleotide diphosphate may interact with the regulatory site of
the ribonucleotide reductase, thereby inhibiting the continuous synthesis of other deoxyribonucleotides (SUHADOLNIK et al., 1967).

Tubercidin (Fig. 61) was isolated from *Streptomyces tubercidius*. Its
physicochemical characteristics are very similar to those of toyocamycin and are summarized by SUHADOLNIK (197o). Tubercidin inhibits
only a few microorganisms, such as *S. faecalis* and *Mycobacterium tuber-*

culosis, but is a potent inhibitor of mouse sarcoma, mouse fibroblasts, human tumors and human KB cells. The cytotoxicity of tubercidin cannot be reversed by purine- or pyrimidine nucleosides. Tubercidin inhibits the growth of the DNA vaccinia virus, of the reovirus
III (double stranded RNA) and mengovirus (single-stranded RNA) (ACS
et al., 1964). Tubercidin affects protein and nucleic acid synthesis
in growing mouse fibroblasts. Tubercidin inhibits the synthesis of
RNA, DNA and protein equally well in cultures of L cells. The antibiotic can be incorporated into RNA - the nucleoside triphosphate
was found within the terminal sequence of tRNA. Tubercidin can also
be incorporated into virus specific RNA and into DNA on reduction to
the corresponding 2'-deoxyribonucleotide. BLOCH et al. (1967) observed that pyruvate prevented the inhibition of the growth of *S. faecalis*
caused by tubercidin and suggested that the impairment of growth is
caused primarily by an interference of tubercidin with the utilization of glucose. But this applies only to bacterial cells because a
mixture of amino acids, nucleosides, ribose-5-phosphate and pyruvate could not prevent the toxic effect of tubercidin on mammalian
cells. In 3T3 cells transformed by SV 4o virus OSSOWSKI and REICH (1972)
observed a strong inhibition of virus-specific RNA by 5-bromotuberci-
din. From hybridization studies with DNA they concluded that 5-bromo-
tubercidin reversibly inhibits both mRNA and rRNA.

Sangivamycin (Fig. 61), the third pyrrolopyrimidine, isolated from
Streptomyces rimosus by RAO and RENN (1963), has very little antibacterial or antifungal activity and inhibits the growth of a limited
number of tumors only. Sangivamycin is phosphorylated by the cell to
the respective mono-, di-, and triphosphates and can be reduced to the
2'-deoxydiphosphate, but to a lesser extent than toyocamycin and tubercidin. Sangivamycin competes for ATP in various polymerization reactions, but not very effectively.

b) Cordycepin

Cordyceptin (Fig. 62) was isolated from *Cordyceps militaris* and *Asper-
gillus nidulans*. Cordycepin, 3'-deoxyadenosine, has cytostatic properties.
Within the cell the compound is phosphorylated to the 5'-mono-, di-,
or triphosphates. In HeLa cells adenosine but no other nucleoside
prevents the inhibitory effect. Cordycepin affects DNA synthesis to
a much lesser degree than RNA synthesis. The antibiotic inhibits the
synthesis of 28S and 18S ribosomal RNA preferentially. In higher organisms the antibiotic is suggested to be a chain terminator since

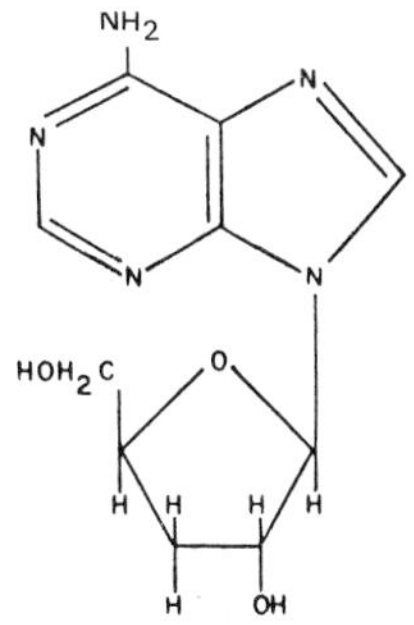

Fig. 62. The structure of cordycepin (3'-deoxyadenosine). (From:
SUHADOLNIK, R.J.: Nucleoside Antibiotics, p. 51. New York-London-
Sidney-Toronto: J. Wiley and Sons, Inc. 197o)

it is incorporated into the growing RNA chain as the last nucleotide (TRUMAN and FREDERIKSEN, 1969).

In mammalian cells, cordycepin has at least three effects on RNA synthesis. The first, complete inhibition, is characteristic of the synthesis of mitochondrial RNA and apparently of mRNA. The second effect, premature termination of nascent RNA and release of the defective RNA chain, occurs in the nucleolus. The synthesis of nuclear heterogeneous RNA seems to be completely insensitive to the drug while that of 5S, 4S and poliovirus RNA is relatively insensitive. PENMAN et al. (197o) speculate that the different responses of RNA synthesis to cordycepin might be explained by the existence of several distinct polymerases with different sensitivities. Also, in isolated nuclei of nerve and glia cells, cordycepin selectively inhibits the formation of ribosomal RNA and its precursors (GRAHN and LØVTRUP-REIN, 1971).

RIZZO et al. (1972) showed that cordycepin is incorporated into the growing RNA chain and terminates chain growth because it lacks the 3'-OH necessary for the formation of the 3',5'-phosphodiester bond. In rat liver cordycepin inhibits the accumulation of ribosoms within 2.5 hours of administration. This effect is due to the premature termination of the transcription of the 45S ribosomal precursor. Cordycepin also inhibits protein synthesis and the corticosteroid-mediated induction of tyrosine transaminase. This is assumed to result from the inhibition of the formation of polyadenylic acid, which is involved in the nuclear processing of messenger RNA.

DARNELL et al. (1971) reported that poly-A synthesis, which seems to be important for the processing of heteregenous RNA and the maturation of cellular mRNA and the maturation of mRNA of DNA viruses could be preferentially blocked by cordycepin. The deoxyadenosine analog has been a useful tool in the study of mRNA of adenoviruses (PHILIPSON et al., 1971).

WU et al. (1972) have investigated if cordycepin also inhibited the replication of RNA tumor viruses. Cordycepin was used to elucidate

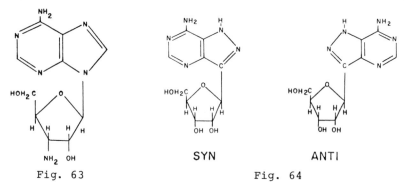

Fig. 63   Fig. 64

Fig. 63. The structure of 3'-amino-3'-deoxyadenosine. (From: SUHADOL-NIK, R.J.: Nucleoside Antibiotics, p. 77. New York-London-Sidney-Toronto: J. Wiley and Sons, Inc. 197o)

Fig. 64. The structure of formycin in the syn and anti conformations. (From: SUHADOLNIK, R.J.: Nucleoside Antibiotics, p. 373. New York-London-Sidney-Toronto: J. Wiley and Sons, Inc. 197o)

the possible role of the poly (A) sequence in virus production induced by 5-iodo-2'-deoxyuridine (JdU) both from seemingly uninfected
murine fibroblasts and from murine Sarcoma virus-transformed non-
producing cells. The results show that cordycepin inhibits the induction of viral production in different inducible strains.

c) 3'-Amino-3'-Deoxyadenosine

This compound (Fig. 63) is an analog of adenosine and a substrate for
adenosine kinase, but not for nucleoside pyrophosphorylase. The antibiotic can be converted to 5-mono-, di-, and triphosphates and is a
potent inhibitor of RNA and DNA synthesis in ascites tumor cells *in
vivo*. This nucleoside inhibits partially purified RNA polymerase from
Micrococcus lysodeicticus and ascites tumor cells; however, DNA-dependent
synthesis of DNA with DNA polymerase I *in vitro* was not inhibited
(SHIGEURA et al., 1966; TRUMAN and KLENOW, 1968).

d) Formycin

Formycin (Fig. 64), in which the pyrrazolo-(4,3-d) pyrimidine ring
has replaced the purine ring, can be taken as an example of the complex interactions which can occur in a cell treated with a nucleoside antibiotic. Formycin is structurally related to adenosine. The
crystal structure and conformation of formycin monohydrates have been
elucidated by circular dichroic spectra as described by PRUSINER et
al. (1973). Formycin effectively replaces adenosine nucleotides in a
number of reactions (WARD and REICH, 1968; WARD et al., 1969). The
enzymes which reportedly can interact with formycin include adenosine
kinase, adenosine deaminase, hexokinase, myokinase, polynucleotide
phosphorylase (*E. coli*), phosphoenolpyruvate kinase, DNA-dependent RNA
polymerase, aminoacyl-tRNA synthetases, tRNA-CCA pyrophosphorylase,
phosphodiesterase, NAD pyrophosphatase and NAD synthetase. Formycin
can further replace adenosine as the terminal nucleoside in a tRNA,
and this tRNA still transfers amino acids to polypeptides.

Formycin B forms a 2:1 complex with polyadenylic acid. The formation
of this complex has been followed by equilibrium dialysis, and by optical rotatory dispersion measurements in the region of 333-35o nm. At
pH 7.o melting curves for thermal dissociation of the complex show a
strongly cooperative helix-coil transition (DAVIES, 1973).

e) 6(p-Hydroxyphenylazo)-Uracil (HPUra)

This compound is synthesized by Imperial Chemical Industries Limited.
BROWN (197o) reported that HPUra inhibits the semi-conservative replication of DNA without affecting DNA repair. Although the compound
selectively and reversibly inhibits DNA replication in *B. subtilis* and
a range of other gram-positive bacteria, it does not stop the replication of phage in *B. subtilis*, nor does it impair the synthesis of
bacterial cell wall materials. To characterize the way in which the
drug acts, *B. subtilis* cells were exposed to lysozyme. DNA synthesis
in this system (BROWN, 197o) was inhibited as in intact bacteria. In
toluene-treated cells, however, DNA replication was much less sensitive to HPUra. If HPUra is a selective inhibitor of DNA semiconser-
vative replication, as BROWN's work indicates, it should prove an
extremely useful tool for identifying the *in vivo* DNA-replicating
enzymes.

B. Mycophenolic Acid

Mycophenolic acid was discovered as early as 1896 by GOSIO and isolated from *Penicillium stoloniferum* cultures. In 1913 ALSBERG and BLACK
named this substance mycophenolic acid. Mycophenolic acid was found
to have antibacterial activity by ABRAHAM (1945), FLOREY et al. (1946)
and GILLIVER (1946), and antifungal activity by FLOREY et al. (1946)
and GILLIVER (1946). Mycophenolic acid inhibits mitosis in mammalian
cells and is active against a variety of experimental tumors in rats
and mice (CARTER,1966; CARTER et al., 1969; and WILLIAMS et al.,
1968).

The chemical structure of mycophenolic acid is presented in Fig. 65.
Many derivatives and analogs of mycophenolic acid have been prepared
(Fig. 66) both chemically and microbiologically, but none is as active
as the natural product. Reduction of the double bond, cyclization to
compounds of types 1 and 2, modification of the phenolic group (for
example, by methylation) or of the aromatic methyl of phthalide methy-
lene groups, lead to destruction of activity. Because simple synthetic
analogs of type 3 are also inactive, it seems that a large part of
the aromatic system is essential for activity.

Mycophenolic acid preferentially inhibits DNA synthesis in L-strains
of fibroblasts *in vitro* (FRANKLIN and COOK, 1969). This inhibition of
DNA synthesis can be reversed by guanine in a non-competitive manner,
but not by hypoxanthine or adenine. The reversal of inhibition by guanine can be suppressed by hypoxanthine, 6-mercaptopurine and adenine.
Therefore it was suggested that the inhibition of nucleic acid synthesis might be due to interference with the biosynthesis of guanine
nucleotides. This suggestion was confirmed in L cells and Landschütz
ascites cells by showing that mycophenolic acid prevented the incorporation of labeled hypoxanthine into xanthine and guanine nucleotides, but not into adenine nucleotides. Preparations of inosinic
acid by dehydrogenase from Landschütz ascites cells or calf thymus

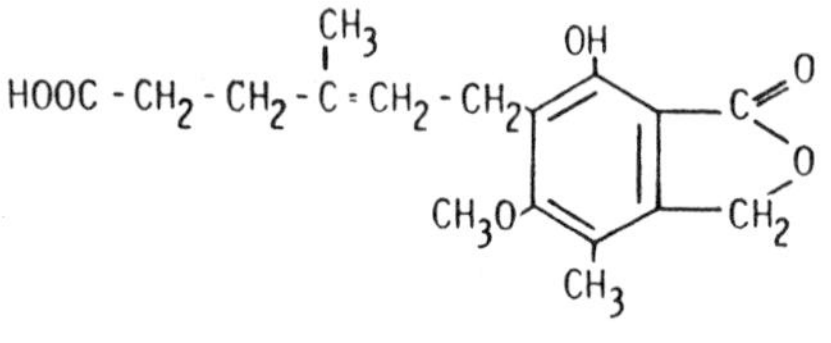

(1)   (2)

Fig. 65 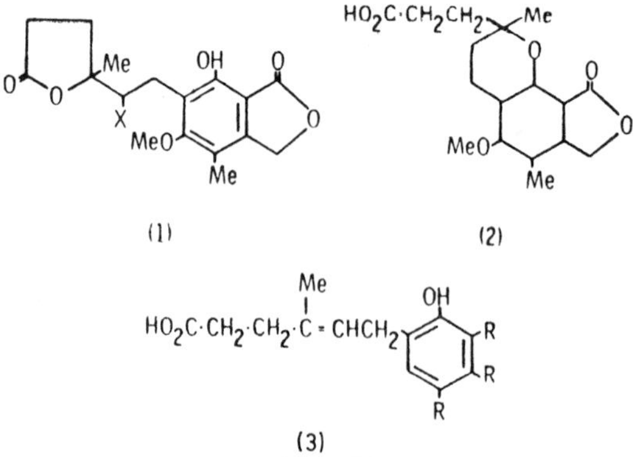   Fig. 66

(3)

Fig. 65. The structure of mycophenolic acid. (According to CARTER
et al., 1969. From: Nature **223**, 849 (1969))

Fig. 66. Derivatives of mycophenolic acid. (According to CARTER et
al., 1969. From: Nature **223**, 849 (1969))

are strongly inhibited by mycophenolic acid. The inhibition showed mixed-type kinetics with K_i values of between 3.o3 x 10^{-8} and 4.5 x 10^{-8} M. Evidence was also obtained for a partial (possibly indirect) inhibition by mycophenolic acid of an early stage of biosynthesis of purine nucleotides, as indicated by a decrease in the accumulation of formylglycine amide ribonucleotide induced by the antibiotic aza-serine in suspensions of Landschütz and Yoshida ascites cells and L cells *in vitro*. It is important in this respect that mycophenolic acid inhibits nucleic acid synthesis at a concentration which suppresses mitosis in those cells. The inhibition of DNA synthesis is not caused by an interaction of this substance with DNA, but probably by inter-action with enzymes involved in the synthesis of DNA precursors (FRANKLIN and COOK, 1969).

Mycophenolic acid has recently become of interest because of its activity against many experimental tumors in rodents and is being investigated in the USA in a wide range of malignancies in man.

C. Amino Acid Analogs

a) Azaserine and DON

Azaserine and DON (6-diazo-5'-oxo-L-norleucine) are produced by Strep-tomycetes and inhibit the growth of several microorganisms; these substances are toxic to mammalian cells and have carcinostatic pro-perties.

The antibiotics are analogs of glutamine (Fig. 67). The structural similarities with glutamine suggest that their mode of action is to interfere competitively with glutamine in the binding of glutamine specific enzymes (BENNET, 1956; HARTMAN et al., 1955; LEVENBERG et al., 1957; TOMISEK et al., 1956; EHRLICH et al., 1956; MOORE and HURLBERT, 1961). Both antibiotics inhibit the synthesis of purine nucleotides, mainly by interfering with the enzyme that is involved in the conversion of 5'-phosphoribosyl-N-formylglycineamide (FGAR) to the corresponding amidine compound, a step in which glutamine serves as donor of amino groups (HENDERSON, 1962).

Glutamine is also involved in the conversion of 5'-phosphoribosyl-1-pyrophosphate to 5'-phosphoribosylamine and in the amination of XMP to GMP. Azaserine and DON antagonize the action of glutamine in a variety of metabolic reactions. However, preferentially they inhibit purine nucleotide biosynthesis. Although azaserine and DON are very similar, important differences exist. DON is significantly more po-tent than azaserine as an inhibitor of purine nucleotide biosynthesis. Azaserine has radiomimetic and mutagenic properties that are lacking in DON.

b) Azotomycin

Azotomycin, closeley related to DON, is a crystalline antibiotic, isolated from culture broths of *Streptomyces ambofaciens* (RAO et al., 1959/196o). Azotomycin inhibits the growth of microorganisms (BROCK-MAN et al., 1969) and of several experimental tumors (ANSFIELD, 1965; CARTER, 1968; WEISS et al., 1968).

Fig. 67 structures:

```
  COOH        COOH        COOH
   |           |           |
 HCNH2       HCNH2       HCNH2
   |           |           |
  CH2         CH2         CH2
   |           |           |
   O          CH2         CH2
   |           |           |
  C = O       C = O       C = O
   |           |           |
  CH          CH          NH2
  ||+         ||+
  N           N
  ||          ||
  N-          N-
```

Azaserine   'DON'   Glutamine

Fig. 67

Azotomycin

```
  CHN2          CH'N2
   |             |
  C = O         C = O          COOH
   |             |              |
  CH2           CH2           CHNH2
   |             |              |
  CH2      |    CH2      |     CH2
   |       |     |       |      |
 CH–NH+CO–CH–NH+CO–CH2
   |       |             |
  COOH     |             |
```

DON   DON   Glutamic acid

Fig. 68

Fig. 67. The structures of azaserine, diazooxonorleucine ('DON') and glutamine. (According to FRANKLIN and SNOW, 1971. From: Biochemistry and Antimicrobial Action. London: Chapman and Hall Ltd. 1971, p. 65) 1971, p. 65)

Fig. 68. The structure of azotomycin. (From: Antimicrobial Agents, 1969, p. 57)

The structure of azotomycin was elucidated by RAO et al. (1959/6o) and identified as a peptide consisting of 1 mole glutamic acid and 2 moles 6-diazo-5-oxo-L-norleucine (DON) (Fig. 68). Enzyme preparations from microorganisms and from tumor cells were shown to hydrolyze azatomycin to DON. The biochemical effects of azotomycin *in vivo* may be accounted for by its hydrolysis to DON.

c) Hadacidin

Hadacidin (N-formyl-hydroxyaminoacetic acid) (Fig. 69) was isolated from many species of Penicillium. Hadacidin is active against a variety of microorganisms. The antibiotic has antitumor properties (GITTER-MAN et al., 1962), but like several other carcinostatic drugs, it has not proved useful for clinical application (HARRIS et al., 1962).

Hadacidin is an analog of L-aspartic acid and able to antagonize the metabolic role of L-aspartic acid in a very specific manner. Hadaci-

```
    H            OH
    |            |
   C = O        C = O
    |            |
   N - OH       HCNH2
    |            |
   CH2          CH2
    |            |
   COOH         COOH
```

Hadacidin   Aspartic Acid

Fig. 69. The structure of hadacidin and aspartic acid. (From: FRANKLIN and SNOW, 1971. London: Chapman and Hall Ltd. 1971, p. 67)

din inhibits the incorporation of various precursors into adenine nucleotides and has, astonishingly, no effect on the biosynthesis of guanine nucleotides. The antibiotic inhibits the conversion of IMP to adenylosuccinic acid, and the inhibition can be partially reversed by L-aspartic acid. These results were obtained with mammalian tumor-cell preparations. Purified adenylosuccinate synthetase from *E. coli* is also sensitive to hadacidin. The antibiotic acts as a competitive inhibitor of L-aspartate. The K_m for L-aspartate in this reaction is 1.5×10^{-4} M (pH 8.o), while the K_i for hadacidin is 4.2×10^{-6} M. Hadacidin is thus an effective competitive antagonist of L-aspartate in the conversion of IMP to adenylosuccinic acid (SHIGEURA and GORDON, 1962). However, other enzymes using L-aspartate as substrate are not inhibited by hadacidin. For example the conversion of 5-amino-4-imi-dazole-N-succinocarboxylic acid ribonucleotide to the corresponding amide which also requires L-aspartate is not affected by hadacidin. The aminoacyl transfer of L-aspartate to the corresponding tRNA is also not inhibited by hadacidin, since hadacidin has no preferential effect on protein biosynthesis. (For review see FRANKLIN and SNOW, 1971).

D. Quinone Antibiotics

Granaticin, Mitomycin Derivatives, Synthetic Quinones

1. Origin, Biological and Chemical Properties

The granaticin antibiotics are produced by several strains of actino-mycetes (CORBAZ et al., 1957). Granaticins are active against gram-positive bacteria (ZÄHNER, personal communication; OGILVIE, 197o; WANKE, 1971).

The chemical structures of granaticin A and granaticin B were eluci-dated by KELLER-SCHIERLEIN et al. (1968). The structure of granati-cin A is shown in Figure 7o. The derivative granaticin B contains a sugar residue attached to the aglycone.

Mitomycin derivatives can be obtained from mitomycins - as described by ROTH et al. (1966). These derivatives, lacking the alkylating aziridine ring, are still active against a variety of gram-positive bacteria and fungi. The structures of derivatives closely related to mitomycin C are shown in Figure 7o.

Synthetic quinones with cytostatic properties are synthesized by Bayer/Leverkusen. These substances inhibit the growth of gram-posi-tive and gram-negative bacteria and have tuberculostatic properties (GRUNDMANN et al., 1969). Quinoline quinones inhibit glycolysis in Ehrlich ascites tumor cells (PÜTTER, 1963). The chemical structure of the compound is shown in Figure 7o.

2. Mechanism of Action

Granaticin, mitomycin derivatives and synthetic quinones have been shown to inhibit the synthesis of RNA and protein in bacteria (KERSTEN

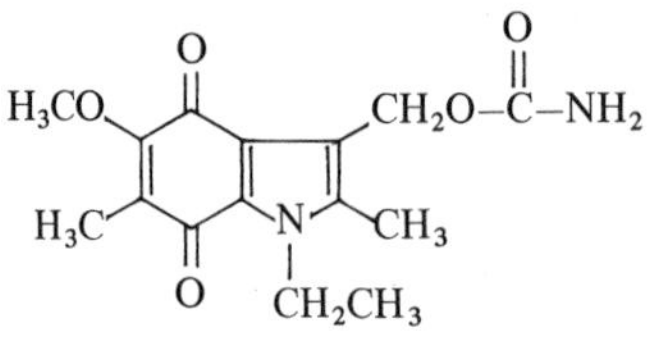

Mitomycin der . I

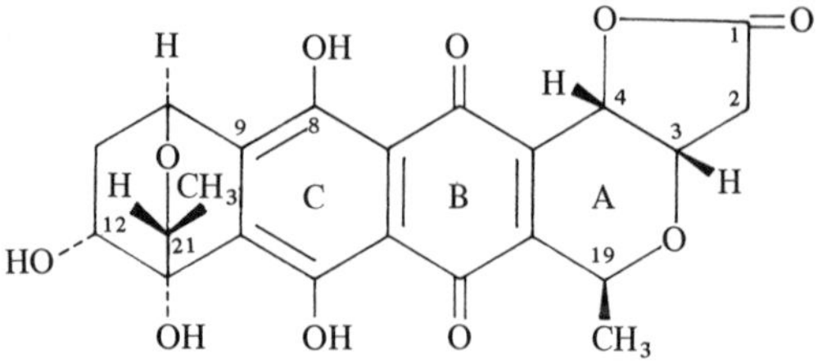

Granaticin A

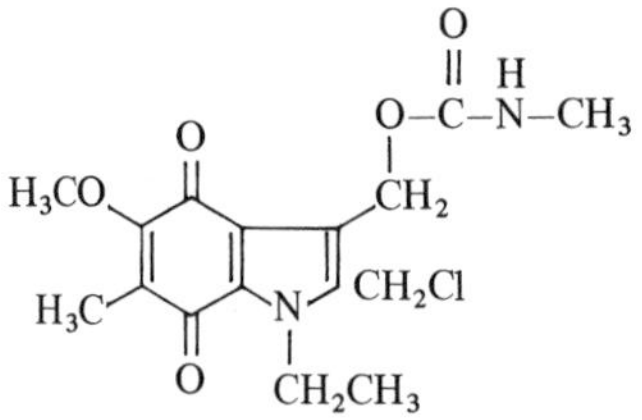

Mitomycin der . II

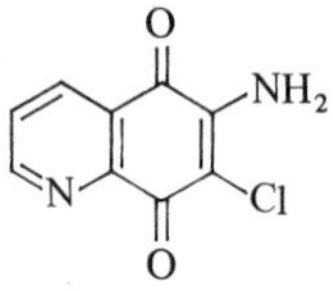

5,8,–Dioxo–6 – amino–7–chloro–
quinoline

Fig. 7o. The structures of quinone antibiotics and a synthetic quinone

and KERSTEN, 1969; KERSTEN et al., 1969, 1970; OGILVIE, 1970; WANKE,
1971; KERSTEN, 1971; OGILVIE et al., 1972).

The effect of quinones on the synthesis of RNA cannot be explained by
an interaction with the DNA template (WANKE et al., 1969) or with the
RNA polymerase (HARTMANN, personal communication; OGILVIE et al.,
1972). It was suggested that these substances might interfere with
the phosphorylation of the RNA precursors (KERSTEN et al., 1969).
Direct evidence that quinones depress the content of GTP and ATP was
obtained (OGILVIE et al., 1972).

The synthetic 5,8-dioxo-6-amino-7-chloroquinoline (abbreviated in the
following as "quinone") was chosen to clarify the mechanism by which
quinones might interfere with RNA synthesis. Incorporation of [^{32}P]
phosphate into nucleotides was measured in E. coli grown in "Tris-
glucose low phosphate medium" in the absence and presence of the "quinone". Parallel to[^{32}P]-labeling, the cells were labeled with ^{3}H-uri-
dine, and the incorporation into RNA was measured over a period of
4o min.

Concentrations of the "quinone" which inhibit net synthesis of RNA
by about 8o% cause a successive decrease in the amount of ^{32}P-labeled
GTP and ATP. The decrease of the ATP pool follows the decrease of the
GTP pool, but with a delay. Concomitantly with the decrease of the
GTP content, the level of guanosine tetraphosphate, guanosine 5'-di-
phosphate, 2'- or 3'-diphosphate (ppGpp) increases (Fig. 71) (OGILVIE
et al., 1973).

It has been postulated that the nucleotide ppGpp plays an important
role in the control of RNA synthesis by amino acids in E. coli strains
with stringent control over RNA synthesis (rel[+] strains) (CASHEL et al.,
1969; CASHEL and KALBACHER, 197o; LAZZARINI et al., 1971). When E. coli rel[+]

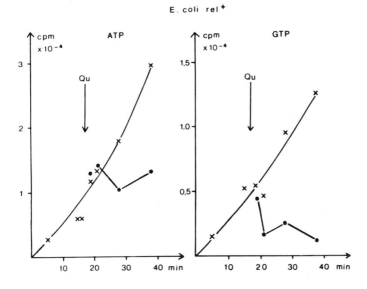

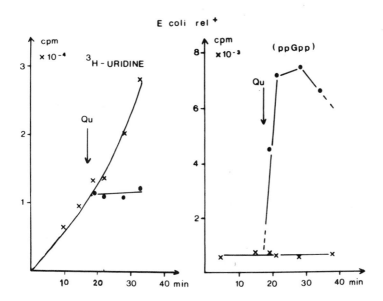

Fig. 71. Effect of the "quinone" on RNA synthesis, ATP, GTP and ppGpp pools in *E. coli* PA 2 (Lavallé). (According to OGILVIE et al., 1972)

cells are deprived of an essential amino acid, or are unable to activate this amino acid. a severe and preferential restriction of ribosomal and tRNA synthesis results. At the same time ppGpp accumulates. *E. coli* rel⁻ strains are unable to control RNA synthesis. In these strains the basal level of ppGpp does not increase appreciably during

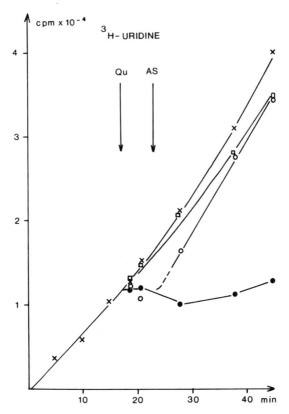

Fig. 72. Reversal of "the quinone" effect on RNA synthesis by a mixture of amino acids. x———x = control; □———□ = control + amino-acids; ●———● = + quinones; O———O = quinones + amino acids. (According to OGILVIE et al., 1972)

amino acid deprivation. Upon re-addition of the required amino acid to deprived cultures of rel+ cells, ppGpp disappears and RNA synthesis resumes.
The inhibitory effect of quinones and quinone antibiotics on RNA synthesis in bacteria can be reversed by the addition of a mixture of amino acids. Several amino-acid mixtures were tested and in all cases it was found that the inhibitory effect could not be abolished by amino acid mixtures lacking leucine, although leucine by itself was ineffective at a concentration of 2 x 1o⁻⁵ M (Fig. 72). In *E. coli* strain PA₂ a combination of leucine and isoleucine reverses the inhibitory effect. The "quinone"-induced increase in the level of ppGpp and the decrease in the levels of ATP and GTP is also abolished by the addition of amino acid mixtures.

Chloramphenicol added to bacteria with stringent control allows RNA synthesis to proceed at a reduced rate of protein synthesis, and this somehow mimics relaxed control. In the presence of chlorampheni-

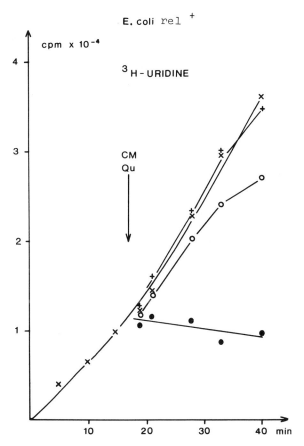

Fig. 73. Reversal of the inhibitory effect of the quinone (Qu) on RNA synthesis by chloramphenicol (CM). x——x = control; +——+ = + chloramphenicol; ●——● = + quinones; ○——○ = + quinones + chloramphenicol. (According to OGILVIE et al., 1972)

col the "quinone" does not inhibit RNA synthesis (Fig. 73). These results suggest that under comparable conditions RNA synthesis in an isogenic *E. coli* strain with relaxed control should have no, or less, effect on RNA synthesis. Experiments with the two isogenic strains of *E. coli* arginine[-] thiamine[-] rel[+] and rel[-] showed that this suggestion was correct. It thus has been established that quinones at low concentrations interfere with the regulation of RNA synthesis in rel[+] bacteria. The quinones interfere with aminoacylation of tRNAs with specificity for these aminoacids.

References

General Introduction

CHIGNELL, Colin F.: Physical Methods for Studying Drug-Protein-Bind-
    ing. In: Hdb. Exp. Pharmakol. Vol. XXVIII Concepts in Biochemical
    Pharmacology (B.B. BRODIE, J.R. GILETTE, Eds.), p. 187-212. Berlin-
    Heidelberg-New York: Springer 1971.
WARING, M.J.: Inhibitors of Nucleic Acid Synthesis. In: GALE, E.F.,
    CUNDLIFFE, E., REYNOLDS, P.E., RICHMOND, M.H., WARING, M.J.: The
    molecular basis of antibiotic action, p. 173-277. London-New York-
    Sydney-Toronto: Wiley 1972.

I. Inhibitors of DNA Synthesis

ALBERTS, B., BARRY, J., HAMANIBA, H., MORAN, L., MACE, D.: Some pro-
    teins of the T_4 bacteriophage, DNA replication apparatus. IUB Ninth
    Int. Congress of Biochemistry, Stockholm 1973, Handbook 3 Sa3.
CAIRNS, J.: DNA synthesis. The Harvey Lectures Series, Vol. 66, p.
    1-18. New York-London: Academic Press 1972.
GANESAN, A.T.: Studies on in vitro replication of B. subtilis DNA. Cold
    Spr. Harb. Symp. quant. Biol. 33, 45-57 (1968).
GANESAN, A.T.: Adenosine triphosphate-dependent synthesis of biolog-
    ically active DNA by azide-poisoned bacteria. Proc. nat. Acad. Sci.
    (Wash.) 68, 1296-13oo (1971).
HOUCK, J.C., IRAUSQUIN, H., LEIKIN, S.: Lymphocyte DNA synthesis in-
    hibition. Science 173, 1139-1141 (1971).
KNIPPERS, R.: DNA polymerase II. Nature 228, 1o5o-1o53 (197o).
KNIPPERS, R., STRÄTLING, W.: The DNA replicating capacity of isolated
    E. coli cell wall-membrane complexes. Nature 226, 713-717 (197o).
KORNBERG, A.: Active center of DNA polymerase. Science 163, 141o-1418
    (1969).
KORNBERG, T., GEFTER, M.L.: DNA synthesis in cell-free extracts of a
    DNA polymerase-defective mutant. Biochem. biophys. Res. Commun. 4o,
    1348-1355 (197o).
MOSES, R.E., RICHARDSON, C.C.: A new polymerase activity of Escherichia
    coli. I. Purification and properties of the activity present in E.
    coli polAl. Biochem. biophys. Res. Commun. 41, 1557-1564 (197o).
OKAZAKI, R., OKAZAKI, T., SAKABE, K., SUGIMOTO, K., KAINUMA, R.,
    SUGINO, A., IWATSUKI, N.: In vivo mechanism of DNA chain growth.
    Cold Spr. Harb. Symp. quant. Biol. 33, 129-143 (1968).
OKAZAKI, R., SUGIMATO, K., OKAZAKI, T., IMAE, Y., SUGINO, A.: DNA
    chain growth: In vivo and in vitro synthesis in a DNA polymerase-
    negative mutant of E. coli. Nature 228, 223-226 (197o).
OKAZAKI, R., SUGINO, A., HIROSE, S., TAMANOI, F.: Discontinous re-
    plication of DNA: Structure and metabolism of RNA-linked DNA frag-
    ments. IUB Ninth Int. Congress of Biochemistry, Stockholm 1973,
    Handbook 3 Sa4.

RICHARDSON, C.C.: Reported in: Wisconsin Symposium on DNA synthesis.
  Nature New Biology 239, 159-16o (1972). (CAMPBELL, J.L., SOLL, L.,
  RICHARDSON, C.C.: Isolation and partial characterization of a mut-
  ant of Escherichia coli deficient in DNA polymerase II. Proc. nat.
  Acad. Sci. (Wash.) 69, 2o9o-2o94 (1972).)
SCHALLER, H., OTTO, B., NÜSSLEIN, V., RUF, J., HERRMANN, R., BONHOEFFER,
  F.: Deoxyribonucleic acid replication in vitro. J. molec. Biol. 63,
  183-2oo (1972).
SMITH, D.W., SCHALLER, H.E., BONHOEFFER, F.J.: DNA Synthesis in vitro.
  Nature 226, 711-713 (197o).

A. Mitomycin

ALBACH, R.A., SHAFFER, J.G.: Effect of mitomycin C metabolism on
  thymidine-methyl-H^3 utilization by Entamoeba histolytica in CLG
  medium. J. Protozool. 14 (Suppl.) 19, 6oa (1967).
ARORA, O.P., SHAH, V.C., RAO, S.R.V.: Studies on micronuclei induced
  by mitomycin C in the root cells of Vicia faba. Expt. Cell Res.
  56, 443-448 (1969).
BASU, S.K., CHAKRABARTY, A.M., ROY, S.C.: Enhancement of catabolite
  repression by mitomycin C in the induced synthesis of β-galactosidase. Biochim. biophys. Acta (Amst.) 1o8, 713-716 (1965).
BEUKERS, R., BERENDS, A.: Isolation and identification of the irradiation product of thymine. Biochim. biophys. Acta (Amst.) 41,
  55o-551 (196o).
BOYCE, R.P., HOWARD-FLANDERS, P.: Genetic control of DNA breakdown
  and repair in E. coli K-12 treated with mitomycin C or ultraviolet
  light. Z. Vererbungsl. 95, 345-35o (1964).
BRUCHOVSKY, N., OWEN, A.A., BECHER, A.J., TILL, J.E.: Effects of
  vinblastine on the proliferative capacity of L cells and their
  progress through the division cycle. Cancer Res. 25, 1232-1237
  (1965).
CARTER, S.K.: Mitomycin C. Cancer Chemother. Rep. Suppl. 1, 99-114
  (1968).
CHEER, S., TCHEN, T.T.: Effect of mitomycin C on the synthesis of
  induced β-galactosidase in E. coli. Biochem. biophys. Res. Comm.
  9, 271-274 (1962).
CHEER, S., TCHEN, T.T.: Effect of mitomycin C on induced enzyme synthesis in E. coli. Bacteriol. Proc. 63, 38 (1963).
COHEN, M.M., SHAW, M.W.: Effects of mitomycin C on human chromosomes.
  J. Cell Biol. 23, 386-395 (1964).
COLES, N.W., GROSS, R.: The effect of mitomycin C on the induced synthesis of penicillinase in Staphylococcus aureus. Biochem. biophys.
  Res. Com. 2o, 366-371 (1965).
CONSTANTOPOULOS, G., TCHEN, T.T.: Enhancement of mitomycin C induced
  breakdown of DNA by inhibitors of protein synthesis. Biochim. biophys. Acta (Amst.) 8o, 456-462 (1964).
COOPER, S., ZINDER, N.D.: The growth of an RNA bacteriophage: The
  role of DNA synthesis. Virology 18, 4o5-411 (1962).
CUMMINGS, D.J.: Macromolecular synthesis during synchronous growth
  of E. coli B/r. Biochim. biophys. Acta (Amst.) 95, 341-35o (1965).
De BOER, D., DIETZ, A., LUMMIS, N.E., SAVAGE, G.M.: Porfiromycin, a
  new antibiotic. I. Discovery and biological activities. In: Antimicrob. Agents Annual 196o, p. 17-22. New York: Plenum Press 1961.
DeWITT, W., HELSINKI, D.R.: Characterization of colicinogenic factor
  E_1 from a non-induced and a mitomycin C-induced Proteus strain.
  J. molec. Biol. 13, 692-7o3 (1965).

DJORDJEVIC, B., KIM, J.H.: Different lethal effects of mitomycin C and actinomycin D during the division cycle of HeLa cells. J. Cell Biol. _38_, 477-482 (1968).

DRISKELL-ZAMENHOF, P.J., ADELBERG, E.A.: Studies on the chemical nature and size of sex factors of E. coli K_{12}. J. molec. Biol. _6_, 483-497 (1963).

FREDERICQ, P.: Colicins. Ann. Rev. Microbiol. 11, 7-22 (1957).

FREESE, E.: Hereditary DNA alterations. Angew. Chemie Int. Ed. _8_, 12-2o (1969).

GERMAN, J., La ROCK, J.: Chromosomal effects of mitomycin, a potential recombinogen in mammalian cell genetics. Tex. Rep. Biol. Med. _27_, 4o9-418 (1969).

GRIBNAU, A.G.M., VELDSTRA, H.: The influence of mitomycin C on the induction of crown-gall tumors. FEBS Letters _3_, 115-117 (1969).

GRULA, E.A., SMITH, G.L., GRULA, M.: Cell division in Erwinia: Inhibition of nuclear body division in filaments grown in penicillin or mitomycin C. Science _161_, 164 (1968).

HANAWALT, P.C.: Cellular recovery from photochemical damage. In: Photophysiology, Vol. 4 (A.C. GIESE, Ed.), p. 2o3-251. New York: Academic Press 1968.

HATA, T., SANO, Y., SUGAWARA, R., MATSUMAE, A., KANAMORI, K., SHIMA, T., HOSHI, T.: Mitomycin, a new antibiotic from Streptomyces. J. Antibiot. (Tokyo) Ser. A _9_, 141-146 (1956).

HATA, T., NOMURA, S., UMEZAWA, I.: Antitumour activity of antibiotic G-253. In: Antimicr. Agents and Chemother. _1966_, 543-545.

HOWARD-FLANDERS, P.: DNA repair. Ann. Rev. Biochem. _37_, 175-2oo (1968).

IIJIMA, T., HAGAWARA, A.: Mutagenic action of mitomycin C on E. coli. Nature _185_, 395-396 (196o).

IONESCO, H., RYTER, A., SCHAEFFER, P.: Sur une bactériophage hérbergé par la souche Marburg de Bacillus subtilis. Ann. Inst. Pasteur 1o7, 764-776 (1964).

IYER, V.N., SZYBALSKI, W.: A molecular mechanism of mitomycin action: Linking of complementary DNA strands. Proc. nat. Acad. Sci. (Wash.) _5o_, 355-362 (1963).

IYER, V.N., SZYBALSKI, W.: Mitomycins and porfiromycins: Chemical mechanism of activation and cross-linking of DNA. Science _145_, 55-58 (1964).

KATO, N., OKABAYASHI, K., MIZUNO, D.: The degradation of ribosomal RNA in E. coli by mitomycin C and AF-5 preferential inhibitors of DNA synthesis. J. Biochem. (Tokyo) _67_, 175-184 (197o).

KERSTEN, H.: Action of mitomycin C on nucleic acid metabolism in tumour and bacterial cells. Biochim. biophys. Acta (Amst.) _55_, 558-56o (1962a).

KERSTEN, H.: Zur Wirkungsweise von Mitomycin C. I. Einfluß von Mitomycin C auf den Desoxyribonucleinsäure-Abbau in ruhenden Bakterien. Hoppe-Seylers Z. physiol. Chem. _329_, 31-39 (1962b).

KERSTEN, H., KERSTEN, W.: Zur Wirkungsweise von Mitomycin C. II. Einfluß von Mitomycin, Chloramphenicol und Mg^{2+} auf den RNA- und DNA-Stoffwechsel in Bakterien. Hoppe-Seylers Z. physiol. Chem. _334_, 141-153 (1963).

KERSTEN, H., KERSTEN, W., LEOPOLD, G., SCHNIEDERS, B.: Effect of mitomycin C on DNAase and RNA in E. coli. Biochim. biophys. Acta (Amst.) _8o_, 521-523 (1964).

KERSTEN, H., KERSTEN, W.: Inhibitors acting on DNA and their use to study DNA replication and repair. In: Inhibitor Tools in Cell Research. (Th. BÜCHNER, H. SIES, Eds.), p. 11-31. Berlin-Heidelberg-New York: Springer 1969.

KERSTEN, H., THEMANN, H.: Morphologische und biochemische Veränderungen an Ascites-Tumorzellen der Maus nach Einwirkung von Mitomycin C. Z. ges. exp. Med. _136_, 2o9-22o (1962).

KIM, J.H., GELBARD, A.S., PEREZ, A.G., EIDINOFF, M.L.: Effect of 5-bromo deoxyuridine on nucleic acid and protein synthesis and viability in HeLa cells. Biochim. biophys. Acta (Amst.) 134, 388-394 (1967).

KIT, S., PIEKARSKI, L.J., DUBBS, D.R.: Effects of 5-fluorouracil, actinomycin D and mitomycin C on the induction of thymidine kinase by vaccinia-infected L-cells. J. molec. Biol. 7, 497-51o (1963).

KNOLLE, P., KAUDEWITZ, F.: Degree of host control on RNA production of an RNA phage. Abstracts, VI. Internat. Congr. Biochem. Vol. 3, p. 234 (1964).

KORN, D., WEISSBACH, A.: Thymineless induction in E. coli $K_{12}\lambda$. Biochim. biophys. Acta (Amst.) 61, 775-79o (1962).

LAWLEY, P.D., BROOKES, P.: Further studies on the alkylation of nucleic acids and their constituent nucleotides. Biochem. J. 89, 127-138 (1963).

LAWLEY, P.D., BROOKES, P.: Interstrand cross-linking of DNA by difunctional alkylating agents. J. molec. Biol. 25, 143-16o (1967).

LEIN, J., HEINEMANN, B., GOUREVITCH, A.: Induction of lysogenic bacteria as a method of detecting potential antitumour agent. Nature 196, 783-784 (1962).

LEMMEL, E.M., GOOD, R.A.: Tolerance of cell mediated immune responses after in vitro treatment of competent cells with mitomycin C. Nature 221, 1164-1165 (1969).

LEOPOLD, G., SCHNIEDERS, B., KERSTEN, H., KERSTEN, W.: The effect of mitomycin C on ribosomes and soluble ribonucleic acid in Escherichia coli. Biochem. Z. 343, 423-432 (1965).

LERMAN, M.I., BENYUMOVICH, M.S.: Effect of mitomycin C on protein synthesis in human neoplastic cell lines. Nature 2o6, 1231-1232 (1965).

LEVINE, M.: Effect of mitomycin C on interactions between temperate phages and bacteria. Virology 13, 493-498 (1961).

LINDQVIST, B., SINSHEIMER, R.L.: The use of mitomycin C as a selective inhibitor of host DNA synthesis in ΦX 174-infected HCr⁻ cells. Fed. Proc. 25, 651 (1966).

LIPSETT, M.N., WEISSBACH, A.: The site of alkylation of nucleic acids by mitomycin. Biochemistry 4, 2o6-211 (1965).

MAGEE, W.E., MILLER, O.V.: Dissociation of the synthesis of host and viral deoxyribonucleic acid. Biochim. biophys. Acta (Amst.) 55, 818-826 (1962).

MATSUMOTO, I., KOZAKA, M., TAKAGI, Y.: Analysis of the acid-soluble deoxyribosidic compounds accumulated in mitomycin C treated bacteria. J. Biochemistry (Tokyo) 6o, 653-659 (1966).

MOORE, G.E., BROSS, I.D.J., AUSMAN, R., NADLER, S., JONES, R., Jr., SLACK, N., RIMM, A.A.: Effects of mitomycin C in 346 patients with advanced cancer. Cancer Chemother. Rep. 52, 675-684 (1968).

MURAKAMI, H.: Electron aspects of the mode of action of the mitomycin molecule. J. theor. Biol. 1o, 236-25o (1966).

NAKATA, Y., NAKATA, K., SAKAMOTO, Y.: On the action mechanism of mitomycin C. Biochem. biophys. Res. Commun. 6, 339-343 (1962).

NATORI, S., HORIGUCHI, T., MIZUNO, D.: Absence of ribonuclease in Alcaligenes faecalis and a possible mechanism of RNA degradation in this bacterium. Biochim. biophys. Acta (Amst.) 134, 337-346 (1967).

NIITANI, H., SUZUKI, A., SHIMOYAMA, M., KIMURA, K.: Effect of mitomycin C injection on lysosomal enzymic activities of Yoshida ascites sarcoma. Gann 55, 447-449 (1964).

NOWELL, P.C.: Mitotic inhibition and chromosome damage by mitomycin in human leucocyte cultures. Exp. Cell Res. 33, 445-449 (1964).

OKAMOTO, K., MUDD, J.A., MANGAN, J., HUANG, W.M., SUBBAIAH, T.V., MARMUR, J.: Properties of the defective phage of B. subtilis. J. molec. Biol. 34, 413-428 (1968a).

OKAMOTO, K., MUDD, J.A., MARMUR, J.: Conversion of B. subtilis DNA
to phage DNA following mitomycin C induction. J. molec. Biol. 34,
429-437 (1968b).
OTSUJI, N.: The effect of glucose on the induction of lambda phage
formation by mitomycin C. Biken's J. 4, 235-241 (1961).
OTSUJI, N.: DNA synthesis and lambda phage development in a lysogenic
strain of E. coli K12. Biken's J. 5, 9-19 (1962).
PAPIRMEISTER, B., DAVISON, C.L.: Unbalanced growth and latent killing
of E. coli following exposure to sulfur mustard. Biochim. biophys.
Acta (Amst.) 103, 70-92 (1965).
PARKIN, J.L., CHIGA, M.: Dissociation of DNA synthesis and mitosis
by mitomycin C in regenerating rat liver. Fed. Proc. 25, 480 (1966).
PATRICK, J.B., WILLIAMS, R.P., MEYER, W.E., FULMOR, W., COSULICH,
D.B., BROSCHARD, R.W, WEBB, J.S.: Aziridinomitosenes: A new class
of antibiotics related to the mitomycins. J. Amer. chem. Soc. 86,
1889-1890 (1964).
RAUTH, A.M.: Evidence for the dark reactivation of mitomycin C anti-
infection damage in mouse cells. Ned. T. Geneesk. 110, 101 (1966).
REICH, E., FRANKLIN, R.M.: Effect of mitomycin C on the growth of
some animal viruses. Proc. nat. Acad. Sci. (Wash.) 47, 1212-1217
(1961).
REINMER, M.V., YOSHIDA, S.: Differential depression of DNA synthesis
in the isolated embryo. Teratology 1, 221 (1968).
ROSS, V.C., SOLYMOSI, I.: Induction of thymidine-kinase in L-132
cells: Dependence of protein synthesis and time of mitomycin action. Fed. Proc. 26, 291 (1967).
ROTT, R., SABER, S., SCHOLTISSEK, C.: Effect on myxovirus of mito-
mycin C, actinomycin D and pretreatment of the host cell with
ultraviolet light. Nature 205, 1187-1190 (1965).
SAKAUCHI, G., DeWITT, C.W.: Immunosuppressive activity of mitomycin
C. Transplantation 5, 248-255 (1967).
SCHWARTZ, H.S., SODERGREN, J.E., PHILIPS, F.S.: Mitomycin C: Chemical
and biological studies on alkylation. Science 142, 1181-1183
(1963).
SEAMAN, E., TARMY, E., MARMUR, J.: Inducible phages of B. subtilis.
Biochemistry 3, 607-613 (1964).
SETLOW, R.B.: The photochemistry, photobiology and repair of poly-
nucleotides. Progr. Nucl. Acid. Res. 8, 257-295 (1968).
SHATKIN, A.J., REICH, E., FRANKLIN, R.M., TATUM, E.L.: Effect of
mitomycin C on mammalian cells in culture. Biochim. biophys. Acta
(Amst.) 55, 277-289 (1962).
SHAW, M.W., COHEN, M.M.: Chromosome exchanges in human leucocytes
induced by mitomycin C. Genetics 51, 181-190 (1965).
SHIBA, S., TERAWAKI, A., TAGUCHI, T., KAWAMATA, J.: Studies on the
effect of mitomycin C on nucleic acid metabolism in E. coli strain
B. Biken's J. 1, 179-193 (1958).
SHIIO, T., WEINBAUM, G., TAKAHASHI, H., MARUO, B.: Chromatographic
analysis of nucleotidic compounds in Bacillus subtilis. J. Gen.
Appl. Microbiol. 8, 178-186 (1962).
SINCLAIR, W.K.: Hydroxyurea: Differential lethal effects on cultured
mammalian cells during the cell cycle. Science 150, 1729-1731
(1965).
SINKUS, A.G.: Effects of mitomycin C on chromosomes in the human cell
culture. Tsitologiya 11, 933-940 (1969).
SMITH-KIELLAND, I.: The effect of mitomycin C on deoxyribonucleic
acid and messenger ribonucleic acid in E. coli. Biochim. biophys.
Acta (Amst.) 114, 254-263 (1966a).
SMITH-KIELLAND, I.: The effect of mitomycin C on ribonucleic acid
synthesis in growing cultures of E. coli. Biochim. biophys. Acta
(Amst.) 119, 486-491 (1966b).

STEIN, G.S., ROTHSTEIN, H.: Mitomycin C may inhibit mitosis by reducing "G_2" RNA synthesis. Curr. Mod. Biology 2, 254-263 (1968).
STICKLER, D.J., TUCKER, R.G., KAY, D.: Bacteriophage-like particles
released from Bacillus subtilis after induction with hydrogen peroxide. Virology 26, 142-145 (1965).
STRAUSS, B.S.: DNA repair mechanism and their relation to mutation
and recombination. Curr. Top. Microbiol. Immunol. 44, 1-85 (1968).
STUDZINSKI, G.P., COHEN, L.S.: Mitomycin C induced increases in the
activities of the deoxyribonucleases of HeLa cells. Biochem.
biophys. Res. Commun. 23, 5o6-512 (1966).
STUDZINSKI, G.P., COHEN, L.S., ROSEMAN, J., SCHWEITZER,J.L.: Elevation
of deoxyribonuclease activities in HeLa cells treated with selective inhibitors of DNA synthesis. Biochem. biophys. Res. Commun.
25, 313-319 (1966).
SUZUKI, H., KILGORE, W.W.: Mitomycin C effect on ribosomes of E. coli.
Science 146, 1585-1587 (1964).
SZYBALSKI, W.: Special microbiological systems. II. Observation on
chemical mutagenesis in microorganisms. Ann. N.Y. Acad. Sci. 76,
475-489 (1958).
SZYBALSKI, W.: Chemical reactivity of chromosomal DNA as related to
mutagenicity: Studies with human cell lines. Cold Spr. Harb. Symp.
quant. Biol. 29, 151-159 (1964).
SZYBALSKI, W., ARNESON, V.G.: Reductive activation and inactivation
of mitomycin as studied with human and bacterial cell cultures.
Molec. Pharmacol. 1, 2o2-2o4 (1965).
SZYBALSKI, W., IYER, V.N.: Cross-linking of DNA by enzymatically or
chemically activated mitomycins and porfiromycins, bifunctionally
"alkylating" antibiotics. Fed. Proc. 23, 946-957 (1964).
SZYBALSKI, W., IYER, V.N.: The mitomycins and porfiromycins. In:
Antibiotics I (D. GOTTLIEB, P.D. SHAW, Eds.), pp. 211-245. Berlin-
Heidelberg-New York: Springer 1967.
TAKAGI, Y.: Action of mitomycin C.: Jap. J. Med. Sci. Biol. 16,
246-249 (1963).
TAKENO, T., NAGATA, T., MIZUNOYA, T.: Photosuppression of mitomycin-
induced lambda-phage development. Nature 218, 295-296 (1968).
TEMIN, H.M., MIZUTANI, S.: RNA-dependent DNA polymerase in virions
of Rous sarcoma virus. Nature 226, 1211-1213 (197o).
TERAWAKI, A., GREENBERG, J.: Post-treatment breakage of mitomycin C
induced cross-links in deoxyribonucleic acid of E. coli. Biochim.
biophys. Acta (Amst.) 119, 54o-546 (1966).
TERESHIN, I.M.: On mechanism of action of mitomycin C on genetic
transformation in hemolytic streptococci. Antibiotiki 14, 796-8oo
(1969).
TOMASZ, M.: Novel assay of 7-alkylation of guanine residues in DNA
application to nitrogen mustard, triethylenemelanine and mitomycin
C. Biochim. biophys. Acta (Amst.) 213, 288-295 (197o).
TSUKAMURA, M., TSUKAMURA, S.: Mutagenic effect of mitomycin C on
Mycobacterium and its combined effect with ultraviolet irradiation.
Jap. J. Microbiol. 6, 53-58 (1962).
VIGIER, P., GOLDE, A.: Action de l'actinomycine D et de la mitomycine
C sur le développment du virus de Rous. C. R. Acad. Sci. (Paris)
258, 389-392 (1964a).
VIGIER, P., GOLDE, A.: Effects of actinomycin D and mitomycin C on
development of Rous sarcoma virus. Virology 23, 511-519 (1964b).
VINCENT, P.C., REEVE, T.S., BRITTLE, N., NICHOLIS, A., RICHARDS, M.:
The effect of cytotoxic drugs on serum albumin in the rat. Aust. J.
expt. Biol. med. Sci. 45, 427-436 (1967).
WACKER, A.: Molecular mechanisms of radiation effects. Progr. Nucl.
Acid. Res. Mol. Biol. 1, 369-399 (1963).

WAKAKI, S., MARUMO, H., TOMIOKA, K., SHIMIZU, G., KATO, E., KAMADA, H., KUDO, S., FUJIMOTO, Y.: Isolation of new fractions of anti-tumour mitomycins. Antibiot. and Chemother. 8, 228-24o (1958).
WEBB, J.S., COSULICH, D.B., MOWAT, J.H., PATRICK, J.B., BROSCHARD, R.W., MEYER, W.E., WILLIAMS, R.P., WOLF, C.F., FULMOR, W., PIDACKS, C., LANCASTER, J.E.: The structures of mitomycin A, B and C, and porfiromycin. Part I. J. Amer. chem. Soc. 84, 3185-3187 (1962).
WEISSBACH, A., LISIO, A.: Alkylation of nucleic acids by mitomycin C and porfiromycin. Biochemistry 4, 196-2oo (1965).
WHITE, H.L., WHITE, J.R.: The binding of porfiromycin to the deoxy-ribonucleic acid. J. Elisha Mitchell Sci. Soc. 81, 37-42 (1965).

B. Streptonigrin

BHUYAN, B.K., Phleomycin, xanthomycin, streptonigrin, nogalamycin and aurantin. In: Antibiotics, I. Mechanism of action. (D. GOTTLIEB, P.D. SHAW, Eds.), pp. 173-18o. Berlin-Heidelberg-New York: Springer 1967.
COHEN, M.M., SHAW, M.W., CRAIG, A.P.: Effects of streptonigrin on cultured human leucocytes. Proc. nat. Acad. Sci. (Wash.) 5o, 16-24 (1963).
HOCHSTEIN, P., LASZLO, J., MILLER, D.: A unique, dicumarol-sensitive, non-phosphorylating oxidation of DPNH and TPNH catalyzed by strep-tonigrin. Biochem. biophys. Res. Commun. 19, 289-295 (1965).
IYER, V.N., SZYBALSKI, W.: Mitomycin and porfiromycin: Chemical mechanism of activation and cross-linking of DNA. Science 145, 55-58 (1964).
LEVINE, M., BOTHWICK, M.: Action of streptonigrin on bacterial DNA metabolism and on induction of phage production in lysogenic bac-teria. Virology 21, 568-579 (1963a).
LEVINE, M., BOTHWICK, M.: Action of streptonigrin on genetic recom-binations between bacteriophages. Proc. XI. Int. Congr. Genet. The Hague/Netherlands 1963b.
MIZUNO, N.S.: Effects of streptonigrin on nucleic acid metabolism of tissue culture cells. Biochim. biophys. Acta (Amst.) 1o8, 394-4o3 (1965).
MIZUNO, N.S., GILBOE, D.P.: Binding of streptonigrin to DNA. Biochim. biophys. Acta (Amst.) 224, 319-327 (197o).
OLESON, J.J., CALDERELLA, L.A., MJOS, K.J., REITH, R.A., THIE, R.S., TOPLIN, I.: The effect of streptonigrin on experimental tumours. Antibiot. and Chemother. 11, 158-164 (1961).
RADDING, C.M.: Uptake of tritiated thymidine by K-12 (λ) induced by streptonigrin. Proc. XI. Int. Congr. Genet., The Hague/Netherlands 1963.
RAO, K.V., CULLEN, W.P.: Streptonigrin, an antitumour substance. I. Isolation and characterization. Antibiotics Annu. 1959/6o, pp. 95o to 953.
RAO, K.V., BIEMANN, K., WOODWARD, R.B.: The structure of streptoni-grin. J. Amer. chem. Soc. 85, 2532-2533 (1963).
WHITE, H.L., WHITE, J.R.: Interaction of streptonigrin with DNA in vitro. Biochim. biophys. Acta (Amst.) 123, 648-651 (1966).
WHITE, H.L., WHITE, J.R.: Lethal action and metabolic effects of streptonigrin on E. Coli. Molec. Pharmacol. 4, 549-565 (1968).
WILSON, W.L., LABRA, C., BARRIST, E.: Preliminary observations on the use of streptonigrin as an antitumour agent in human beings. Antibiotic. and Chemother. 11, 147-15o (1961).

C. Sibiromycin

BRAZHNIKOVA, M.G., KOVSHAROVA, I.N., KONSTANTINOVA, N.V., MESENTSEV, A.S., PROSHLJAKOVA, V.V., TOLSTYCH, I.V.: Chemical study on anti-tumour antibiotic sibiromycin. Antibiotiki 15, 297-3oo (197o).

BRAZHNIKOVA, M.G., KONSTANTINOVA, N.V., MESENTSEV, A.S.: Sibiromycin: Isolation and characterization. J. Antibiot. (Tokyo) 25, 668-673 (1972).

DUDNIK, Y.V., NETYKSA, E.M., VARIK, O.Y.: Increased antibacteria effect of bruneomycin and sibiromycin in cultures with impaired reparation of DNA. Antibiotiki 16, 487-491 (1971a).

DUDNIK, Y.V., KARPOV, V.L., NETYKSA, E.M.: Sulphur-containing derivation of sibiromycin. Removal of sulphur on interaction with DNA. Antibiotiki 16, 6-8 (1971b).

GAUSE, G.F., DUDNIK, Y.V.: Interaction of antitumour antibiotics with DNA: Studies on sibiromycin. Progr. Molec. Subcell. Biology 2, 33-39 (1969); Berlin-Heidelberg-New York: Springer 1969.

GAUSE, G.F., DUDNIK, Y.V., NETYKSA, E.M., LAIKO, A.V., GAUSE, G.G.: Mechanism of action of sibiromycin. Antibiotiki 15, 867-871 (197o).

GAUSE, G.G., DUDNIK, Y.V., DOLGILEVICH, S.M.: Suppression of nucleic acid synthesis by sibiromycin. Antibiotiki 17, 413-419 (1972).

D. Phleomycin and E. Bleomycin

ARGOUDELIS, A.D., BERGY, M.E., PYKE, T.R.: Zorbamycin and related antibiotics. I. Production, isolation and characterization. J. Antibiot. (Tokyo) 24, 543-557 (1971).

BARRANCO, S.C., HUMPHREY, R.M.: The effect of bleomycin on survival and cell progression on Chinese hamster cells in vitro. Cancer Res. 31, 1218-1223 (1971).

BRADNER, W.T., PINDELL, H.M.: Antitumour properties of phleomycin. Nature 196, 682-683 (1962).

BRADNER, W.T., PINDELL, M.H.: Strain specificity of stimulated regression of sarcoma 18o. Cancer Res. 25, 859-564 (1965).

DJORDJEVIC, B., KIM, J.H.: Lethal effect of phleomycin in different stages of the division cycle of HeLa cells. Cancer Res. 27, 2255 to 226o (1967).

ENDO, H.: Qualitative difference between bleomycin and radiation effects on cell viability. J. Antibiot. (Tokyo) 23, 5o8-51o (197o).

FALASCHI, A., KORNBERG, A.: Phleomycin, an inhibitor of DNA polymerase. Fed. Proc. 23, 94o-945 (1964).

FUJIWARA, Y., KONDO, T.: Strand-scission of HeLa cell deoxyribonucleic acid by bleomycin in vitro and in vivo. Biochem. Pharmacol. 22, 323-333 (1973).

GORMAN, T., PIETSCH, P.: Strategy for crystallographic analysis of phleomycin-DNA complexes: Fourier transforms. Physiol. Chem. Phys. 1, 312-316 (1969).

GRIGG, G.W.: Induction of DNA breakdown and death in E. coli by phleomycin. Its association with dark-repair processes. Mol. gen. Genet. 1o4, 1-11 (1969).

GRIGG, G.W.: Amplification of phleomycin induced death and DNA breakdown by caffeine in E. coli. Molec. gen. Genet. 1o7, 162-172 (197o).

HAIDLE, C.W.: Fragmentation of deoxynucleic acid by bleomycin. Molec. Pharmacol. 7, 645-652 (1971).

HAIDLE, C.W., WEISS, K.K., MACE, M.L., Jr.: Induction of bacteriophage by bleomycin. Biochem. biophys. Res. Commun. 48, 1179-1184
(1972).

HECHT, T., SUMMERS, D.F.: Effect of phleomycin on poliovirus RNA
replication. Virology 4o, 441-447 (197o).

HOTTA, Y., STERN, H.: Action of phleomycin on meiotic cells. Cancer
Res. 29, 1699-17o6 (1969).

IKEKAWA, T., IWAMI, F., HIRANAKA, H., UMEZAWA, H.: Separation of
phleomycin components and their properties. J. Antibiot. (Tokyo)
Ser. A 17, 194-199 (1964).

ISHIZUKA, M., TAKAYAMA, H., TAKEUCHI, T., UMEZAWA, H.: Studies on
antitumour activity, antimicrobial activity and toxicity of phleomycin. J. Antibiot. (Tokyo) Ser. A 19, 26o-271 (1966).

ISHIZUKA, M., TAKAYAMA, H., TAKEUCHI, T., UMEZAWA, H.: Activity and
toxicity of bleomycin. J. Antibiot.(Tokyo) Ser. A 2o, 15-24 (1967).

ITO, Y., OHASHI, Y., EGAWA, Y., YAMAGUCHI, T., FURUMAI, T., ENOMOTO,
K., OKUDA, T.: Antibiotica YA 56, a new family of phleomycin-bleo-
mycin group antibiotics. J. Antibiot. (Tokyo) 24, 727-731 (1971).

IWATA, A., CONSIGLI, R.H.: Effect of phleomycin on polyoma virus
synthesis in mouse embryo cells. J. Virol. 7, 29-4o (1971).

KAJIWARA, K., KIM, U.H., MUELLER, G.C.: Phleomycin, an inhibitor of
replication of HeLa cells. Cancer Res. 26, 233-236 (1966).

KATZ, S.: The reversible reaction of Hg(II) and double-stranded
polynucleotides. A step-function theory and its significance. Biochim. biophys. Acta (Amst.) 68, 25o-253 (1963).

KIHLMAN, B.A., ODMARK, G., HARTLEY, B.: Studies on the effects of
phleomycin on chromosome structure and nucleic acid synthesis in
Vicia faba. Mutation Res. 4, 783-79o (1967).

KOCH, G.: Differential effect of phleomycin on the infectivity of
poliovirus and poliovirus-induced ribonucleic acids. J. Virol.
8, 28-34 (1971).

KOYAMA, G., NAKAMURA, H., MURAOKA, Y., TAKITA, T., MAEDA, K., UMEZAWA,
H.: The chemistry of bleomycin. II. The molecular and crystal struc-
ture of a sulfur-containing chromophoric amino acid. Tetrahedron
Letters pp. 4635-4638 (1968).

KRUEGER, W.C., PSCHIGODA, L.M., REUSSER, F.: Interactions of DNA
with zorbamycin, phleomycin and bleomycin; ultraviolet absorption
and circular dichroism measurements. J. Antibiot. (Tokyo) 26,
424-428 (1973).

MAEDA, K., KOSAKA,H., YAGISHITA, K., UMEZAWA, H.: A new antibiotic,
phleomycin. J. Antibiot. (Tokyo) Ser. A 9, 82-85 (1956).

MATTINGLY, E.: Induction of chromosome and chromatid-type aberrations
by phleomycin. Mutation Res. 4, 51-57 (1966).

MIYAKI, M., ONO, T., UMEZAWA, H.: Inhibition of ligase reaction by
bleomycin. J. Antibiot. (Tokyo) 24, 587-592 (1971).

MÜLLER, W.E.G., YAMAZAKI, Z., ZAHN, R.K.: Bleomycin, a selective inhibitor of DNA-dependent DNA polymerase from oncogenic RNA virus.
Biochem. biophys. Res. Commun. 46, 1667-1673 (1972a).

MÜLLER, W.E.G., YAMAZAKI, Z., BRETER, H.J., ZAHN, R.K.: Action of
bleomycin on DNA and RNA. Europ. J. Biochem. 31, 518-525 (1972b).

MÜLLER, W.E.G., YAMAZAKI, Z., ZÖLLNER, J.E., ZAHN, R.K.: Action of
bleomycin on programmed synthesis. Influence on DNA and RNA nucleases. FEBS Letters 31, 217-221 (1973).

MURAOKA, Y., TAKITA, T., MAEDA, K., UMEZAWA, H.: Chemistry of bleomycin. IV. The structure of amino component II of bleomycin A2.
J. Antibiot. (Tokyo) 23, 252-253 (197o).

MURAOKA, Y., TAKITA, T., MAEDA, K., UMEZAWA, H.: Chemistry of bleomycin. VI: Selective cleavage of bleomycin A2 by N-bromosuccinimide.
J. Antibiot. (Tokyo) 25, 185-186 (1972).

NAGAI, K., SUZUKI, H., TANAKA, N., UMEZAWA, H.: Decrease of melting temperature and single strand scission of DNA by bleomycin in the presence of hydrogen peroxide. J. Antibiot. (Tokyo) 22, 624-628 (1969a).

NAGAI, K., YAMAKI, H., SUZUKI, H., TANAKA, N., UMEZAWA, H.: The combined effects of bleomycin and sulfhydryl compounds on the thermal denaturation of DNA. Biochim. biophys. Acta (Amst.) 179, 165 to 171 (1969b).

OHKI, M., TOMIZAWA, J.-I.: Assymetric transfer of DNA strands in bacterial conjugation. Cold Spr. Harb. Symp. quant. Biol. 33, 651-658 (1968).

OMOTO, S., TAKITA, T., MAEDA, K., UMEZAWA, H., UMEZAWA, S.: The chemistry of bleomycin. VIII. The structure of the sugar moiety of bleomycin A_2. J. Antibiot. (Tokyo) 25, 752-754 (1972).

PIETSCH, P.: Reactions of phleomycin and DNA. J. Cell Biol. 31, 68A to 87A (1966).

PIETSCH, P.: Differences in DNA synthesis as reflected in variations in the acute inhibition of replication by the antibiotic phleomycin. Anat. Rec. 157, 3o1-3o1 (1967).

PIETSCH, P.: Structural events in DNA in transcription and replication: the influence of histones on in vitro reactions of actinomycin-D and phleomycin-9o9. Cytobios 1, 375-391 (1969).

PIETSCH, P., CORBETT, C.: Competitive effects of phleomycin and mercuric chloride in vivo. Nature 219, 933-934 (1968).

PIETSCH, P., GARRETT, H.: Primary site of reaction in the in vitro complex of phleomycin in DNA. Nature 219, 488-489 (1968).

PIETSCH, P., GARRETT, H.: Phleomycin: evidence of in vivo binding to DNA. Cytobios 1, 7-15 (1969a).

PIETSCH, P., GARRETT, H.: Phleomycin induced changes in the ultrastructure of DNA. Biophys. J. 9A, 126 (1969b).

PIETSCH, P., McCOLLISTER, S.B.: Replication and the activation of muscle differentiation. Nature 2o8, 117o-1173 (1965).

PITTS, J., SINSHEIMER, R.L.: Effect of phleomycin on replication of bacteriophage Φ X 174. J. molec. Biol. 15, 676-68o (1966).

SAUNDERS, P.P., SCHULTZ, G.A.: Mechanism of action of bleomycin-I. Bacterial growth studies. Biochem. Pharmacol. 21, 1657-1666 (1972).

SHIRAKAWA, I., AZEGAMI, M., ISHII, S., UMEZAWA, H.: Reaction of bleomycin with DNA. Strand scission of DNA in the absence of sulfhydryl or peroxide compounds. J. Antibiot. (Tokyo) 24, 761-766 (1971).

SUZUKI, H., NAGAI, K., AKUTSU, E., YAMAKI, H., TANAKA, N., UMEZAWA, H.: On the mechanism of action of bleomycin. Strand scission of DNA caused by bleomycin and its binding to DNA in vitro. J. Antibiot. (Tokyo) 23, 473-48o (197o).

TAKITA, T.: Studies on purification and properties of phleomycin. J. Antibiot. (Tokyo) Ser. A 12, 285-289 (1959).

TAKITA, T., MAEDA, K., UMEZAWA, H., OMOTO, S., UMEZAWA, S.: Chemistry of bleomycin. III. The sugar moieties of bleomycin A_2. J. Antibiot. (Tokyo) Ser. A 22, 237-239 (1969).

TAKITA, T., MURAOKA, Y., MAEDA, K., UMEZAWA, H.: Chemical studies on bleomycin. I. The acid hydrolysis products of bleomycin A_2. J. Antibiot. (Tokyo) 21, 79-8o (1968).

TAKITA, T., MURAOKA, Y., OMOTO, S., KOYAMA, G., FUJII, A., MAEDA, K., UMEZAWA, H.: Chemical studies on an antitumour antibiotic, bleomycin A_2. In: Progr. Antimicrob. Agent and Anticancer Chemotherapy, Vol. II, pp. 1o31-1o36. Baltimore/Maryland-Manchester/England: University Park Press 197o.

TAKITA, T., MURAOKA, Y., FUJII, A., ITOH, H., MAEDA, K., UMEZAWA, H.: The structure of the sulfur-containing chromophore of phleomycin, and chemical transformation of phleomycin to bleomycin. J. Antibiot. (Tokyo) 25, 197-199 (1972a).

TAKITA, T., MURAOKA, Y., YOSHIOKA, T., FUJII, A., MAEDA, K., UMEZAWA,
  H.: The chemistry of bleomycin. IX. The structures of bleomycin
  and phleomycin. J. Antibiot. (Tokyo) 25, 755-758 (1972b).
TAKITA, T., YOSHIOKA, T., MURAOKA, Y., MAEDA, K., UMEZAWA, H.:
  Chemistry of bleomycin. V. Revised structure of an amine component
  of bleomycin A₂. J. Antibiot. (Tokyo) 24, 795-796 (1971).
TANAKA, N., YAMAGUCHI, H., UMEZAWA, H.: Mechanism of action of phleo-
  mycin, a tumour-inhibitory antibiotic. Biochem. biophys. Res.
  Commun. 1o, 171-174 (1963).
TERASIMA, T., UMEZAWA, H.: Lethal effect of bleomycin on cultured
  mammalian cells. J. Antibiot. (Tokyo) 23, 3oo-3o4 (197o).
TERASIMA, T., YASUKAWA, M., UMEZAWA, H.: Breaks and rejoining of DNA
  in cultured mammalian cells treated with bleomycin. Gann 61, 513
  to 516·(197o).
UMEZAWA, H., MAEDA, K., TAKEUCHI, T., OKAMI, Y.: New antibiotics,
  bleomycin A and B. J. Antibiot. (Tokyo) 19A, 2oo-2o9 (1966a).
UMEZAWA, H., SUHARA, Y., TAKITA, T., MAEDA, K.: Purification of
  bleomycins. J. Antibiot. (Tokyo) 19A, 21o-215 (1966b).
WATANABE, M., AUGUST, J.T.: Inhibition of RNA synthesis by phleomycin.
  Bact. Proc. 66, 115 (1966).
WATANABE, M., AUGUST, J.T.: Replication of RNA bacteriophage R23.
  II. Inhibition of phage-specific RNA synthesis by phleomycin.
  J. molec. Biol. 33, 21-33 (1968).
YAMANE, T., DAVIDSON, N.: On the complexing of deoxyribonucleic acid
  (DNA) by mercuric ion. J. Amer. chem. Soc. 83, 2599-26o7 (196o).
YAMAZAKI, Z., MÜLLER, W.E.G., ZAHN, R.K.: Action of bleomycin on
  programmed synthesis. Influence on enzymatic DNA, RNA and protein
  synthesis. Biochim. biophys. Acta (Amst.) 3o8, 412-421 (1973).
ZEE-CHENG, K.Y., CHENG, C.: Synthesis of 2'-(2-aminoethyl)-2',4'-
  bithiazole-4-carboxylic acid, a component of the antitumour antibiotic bleomycin. J. heterocyclic Chem. 7, 1439-144o (197o).

F. Neocarcinostatin

FALASCHI, A., KORNBERG, A.: Phleomycin, an inhibitor of DNA polymer-
  ase. Fed. Proc. 23, 94o-945 (1964).
HEINEMANN, B., HOWARD, A.J.: Induction of lambda bacteriophage in
  E. coli as a screening test for potential antitumour agents.
  Appl. Microbiol. 12, 234-239 (1964).
HOMMA, M., KOIDA, T., SAITO-KOIDE, T., KAMO, I., SETO, M., KUMAGAI,
  K., ISHIDA, N.: Specific inhibition of the initiation of DNA
  synthesis in HeLa cells by neocarzinostatin. In: Progress in Anti-
  microb. and Anticancer Chemotherapy, Vol. 2, pp. 41o-415. Baltimore/
  Maryland-Manchester/England: University Park Press 197o.
ISHIDA, N., MIYAZAKI, K., KUMAGI, K., RIKIMARU, M.: Neocarzinostatin,
  an antitumour antibiotic of high molecular weight. Isolation, phy-
  sicochemical properties and biological activities. J. Antibiot.
  (Tokyo) Ser. A 18a, 68-76 (1965).
KAJIWARA, K., KIM, U.H., MÜLLER, G.C.: Phleomycin, an inhibitor of
  replication of HeLa cells. Cancer Res. 26, 233-236 (1966).
KUMAGAI, K., ONO, Y., NISHIKAWA, T., ISHIDA, N.: Cytological studies
  on the effect of neocarzinostatin on HeLa cells. J. Antibiot.
  (Tokyo) Ser. A 19, 69-74 (1966).
MAEDA, H., ISHIDA, N.: Conformational study of antitumour proteins.
  Neocarzinostatin and a deaminated derivative. Biochim. biophys.
  Acta (Amst.) 147, 597-599 (1967).
MAEDA, H., KUMAGAI, K., ISHIDA, N.: Characterization of neocarzino-
  statin. J. Antibiot. (Tokyo) Ser. A 19, 253-259 (1966).

MEIENHOFER, J., MAEDA, H., GLASER, Ch.B., CZOMBOS, J., KUROMIZU, K.:
  Primary structure of neocarzinostatin, an antitumour protein.
  Science 178, 875-876 (1972).
ONO, Y., WATANABE, Y., ISHIDA, N.: Mode of action of neocarzinostatin.
  Inhibition of DNA synthesis and degradation of DNA in Sarcina lutea.
  Biochim. biophys. Acta (Amst.) 119, 46-58 (1966).
PRICE, K.E., BRUCK, R.E., LEIN, J.: System for detecting inducers of
  lysogenic E. coli W 17o9 (λ) and its applicability as a screen for
  antineoplastic antibiotics. Appl. Microbiol. 12, 428-435 (1964).

G. Edeine

HETTINGER, T.P., KURYLO-BOROWSKA, Z., CRAIG, L.C.: The chemistry of
  the edeine polyamine antibiotics. Ann. N.J. Acad. Sci. 171, 1oo2
  to 1oo9 (197o).
HETTINGER, T.P., CRAIG, L.C.: Edeine. IV. Structures of the antibiotic peptides edeines A₁ and B₁. Biochemistry 9, 1224-1232 (197o).
KURYLO-BOROWSKA, Z., SZER, W.: Inhibition of bacterial DNA synthesis
  by edeine effect on E. coli mutants lacking DNA polymerase I.
  Biochim. biophys. Acta (Amst.) 287, 236-245 (1972).
TABACZYNSKI, M.M., JABLONSKA, E.: The effect of edeine on intracellular growth and mutation of bacteriophage T4 B. Acta Microbiol. pol.
  2(A), 169-178 (197o).
WOJCIECHOWSKA, H., CIARKOWSKI, J., CHMARA, H., BOROWSKI, E.: The
  antibiotic edeine IX: The isolation and the composition of edeine D.
  Experientia 28, 1423-1424 (1972).

H. Nalidixic Acid

BAIRD, J.P., BOURGUIGNON, G.J., STERNGLANZ, R.: Effect of nalidixic
  acid on the growth of deoxyribonucleic acid bacteriophages. J.
  Virol. 9, 17-21 (1972).
BARBOUR, S.D.: Effect of nalidixic acid on conjugational transfer and
  expression of episomal lac genes in Escherichia coli K12. J. molec.
  Biol. 28, 373-376 (1967).
BONHOEFFER, F., VIELMETTER, W.: Conjugational DNA transfer in Escherichia coli. Cold Spr. Harb. quant. Biol. 33, 623-627 (1968).
BOUCK, N., ADELBERG, E.A.: Mechanism of action of nalidixic acid on
  conjugating bacteria. J. Bact. 1o2, 688-7o1 (197o).
BOYLE, J.V., COOK, T.M., GOSS, W.A.: Mechanism of action of nalidixic
  acid on E. coli. VI. Cell free studies. J. Bact. 97, 23o-236
  (1969).
COOK, T.M., DEITZ, W.H., GOSS, W.A.: Mechanism of action of nalidixic
  acid on E. coli. IV. Effects on the stability of cellular constituents. J. Bact. 91, 774-779 (1966a).
COOK, T.M., GOSS, W.A., DEITZ, W.H.: Mechanism of action of nalidixic
  acid on E. coli. V. Possible mutagenic effect. J. Bact. 91, 78o-783
  (1966b).
COOK, T.M., BROWN, K.G., BOYLE, J.V., GOSS, W.A.: Bactericidal action
  of nalidixic acid on B. subtilis. J. Bact. 92, 151o-1514 (1966c).
DEITZ, W.H., COOK, T.M., GOSS, W.A.: Mechanism of action of nalidixic
  acid on E. coli. III. Conditions required for lethality. J. Bact.
  91, 768-773 (1966).

GAGE, L.P., FUJITA, D.J.: Effect of nalidixic acid on deoxyribonucleic acid synthesis in bacteriophage SPO1-infected Bacillus subtilis. J. Bact. 98, 96-1o3 (1969).

GANESAN, A.T.: Studies on the in vivo synthesis of transforming DNA. Proc. nat. Acad. Sci. (Wash.) 61, 1o58-1o65 (1968).

GOSS, W.A., DEITZ, W.H., COOK, T.M.: Mechanism of action of nalidixic acid on E. coli. J. Bact. 88, 1112-1118 (1964).

GOSS, W.A., DEITZ, W.H., COOK, T.M.: Mechanism of action of nalidixic acid on E. coli. II. Inhibition of DNA synthesis. J. Bact. 89, 1o68-1o74 (1965).

HANE, M.W.: Some effects of nalidixic acid on conjugation in Escherichia coli K12. J. Bact. 1o5, 46-56 (1971).

HANE, M.W., WOOD, T.H.: E. coli K-12 mutants resistant to nalidixic acid: Genetic mapping and dominance studies. J. Bact. 99, 238-241 (1969).

LESHER, G.Y., FROELICH, E.J., GRUETT, M.D., BAILEY, J.H., BRUNDAGE, R.P.: 1,8-naphthyridine derivatives: a new class of chemotherapeutic agents. J. med. pharm. Chem. 5, 1o63-1o65 (1962).

MOUNOLOU, J.C., PERRODIN, G.: Inhibition de l'adaptation respiratoire et de la synthese d'ADN par l'acide nalidixique. Compt. Rend. Acad. Sci. (Paris) 267 D, 1286-1288 (1968).

OHKI, M., THOMIZAWA, J.-I.: Assymetric transfer of DNA strands in bacterial conjugation. Cold Spr. Harb. Symp. quant. Biol. 33, 651 to 658 (1968).

PEDRINI, A.M., GEROLDI, D., FALASCHI, A.: Nalidixic acid does not inhibit bacterial transformation. Molec. gen. Genet. 116, 91-94 (1972a).

PEDRINI, A.M., GEROLDI, D., SICCARDI, A., FALASCHI, A.: Studies on the mode of action of nalidixic acid. Europ. J. Biochem. 25, 359 to 365 (1972b).

PUGA, A., TESSMAN, I.: Mechanism of transcription of bacteriophage S13. II. Inhibition of phage-specific transcription by nalidixic acid. J. molec. Biol. 75, 99-1o8 (1973).

RAMAREDDY, G., REITER, H.: Specific loss of newly replicated deoxyribonucleic acid in nalidixic acid-treated B. subtilis 168. J. Bact. 1oo, 724-729 (1969).

ROSENKRANZ, H.S., LAMBEK, C.: In vivo effect of nalidixic acid (Neg Gram) on the DNA of human diploid cells in tissue culture. Proc. Soc. exp. Biol. (N.Y.) 12o, 549-552 (1965).

SCHNECK, P.K., STAUDENBAUER, W.L., HOFSCHNEIDER, P.H.: Replication of bacteriophage M-13. Template specific inhibiton of DNA synthesis by nalidixic acid. Europ. J. Biochem. 38, 13o-136 (1973).

WALTON, J.R., SMITH, D.H.: Hemolysin production in E. coli associated with nalidixic acid resistance. Antimicrob. Agents Chemother. 1968, pp. 54-56 (1968).

WEHR, C.T., KUDRNA, R.D., PARKS, L.W.: Effect of putative DNA inhibitors on macromolecular synthesis in S. cerevisiae. J. Bact. 1o2, 636-641 (197o).

II. Inhibitors of RNA Synthesis that Interact with the DNA Template

A. Actinomycin

ACS, G., REICH, E., VALANJU, S.: Ribonucleic acid metabolism of Bacillus subtilis. Effects of actinomycin. Biochim. biophys. Acta (Amst.) 76, 68-79 (1963).

ANGERMAN, N.S., VICTOR, .T.A., BELL, C.L., DANYLUK, S.S.: A proton magnetic resonance study of the aggregation of actinomycin D in D_2O. Biochemistry 11, 24o2-2411 (1972).

ARISON, B.H., HOOGSTEEN, K.: Nuclear magnetic resonance spectral studies on actinomycin D. Preliminary observations on the effect of complex formation with S'-deoxyguanylic acid. Biochemistry 9, 3976-3983 (197o).

ASCOLI, F., De SANTIS, P., SAVINO, M.: Solvent effect on the conformation of actinomycin D. Nature 227, 1237-1241 (197o).

BADER, J.P.: The role of deoxyribonucleic acid in the synthesis of Rous sarcoma virus. Virology 22, 462-468 (1964).

BALTIMORE, D.: Viral RNA-dependent DNA polymerase. Nature 226, 12o9-1211 (197o).

BARRY, R.D.: The effects of actinomycin D and ultraviolet irradiation on the production of fowl plague virus. Virology 24, 563-569 (1964).

BARRY, R.D., IVES, D.R., CRUICKSHANK, J.G.: Participation of deoxyribonucleic acid in the multiplication of influenza virus. Nature 194, 1139-114o (1962).

BASERGA, R., ESTENSEN, R.D., PETERSEN, R.O., LAYDE, J.P.: Inhibition of DNA synthesis in Ehrlich ascites cells by actinomycin D. I. Delayed inhibition by low doses. Proc. nat. Acad. Sci. (Wash.) 54, 745-751 (1965).

BEABEALASHVILLY, R.S., GURSKY, G.V., SAVOTCHKINA, L.P., ZASEDATELEV, A.S.: RNA polymerase-DNA complexes. III. Binding of actinomycin D to RNA polymerase-DNA complex. Biochim. biophys. Acta 294, 425 to 433 (1973).

BENEDETTO, A., DELFINI, C., PULEDDA, S., SEBASTIANI, A.: Actinomycin D binding to $_{37}$RC and HeLa cell lines. Biochim. biophys. Acta (Amst.) 287, 33o-339 (1972).

BERNARDI, G.: Mechanism of action and structure of acid deoxyribonuclease. Adv. Enzymol. 31, 1-49 (1968).

BLOBEL, G., POTTER, V.R.: Studies on free and membrane-bound ribosomes in rat liver. II. Interaction of ribosomes and membranes. J. molec. Biol. 26, 293-3o1 (1967).

BROCKMANN, H.: Structural differences of the actinomycins and their derivatives. Ann. N.Y. Acad. Sci. 89, 323-335 (196oa).

BROCKMANN, H.: Die Actinomycine. Fortschr. Chem. org. Naturstoffe 18, 1-54 (196ob).

BROCKMANN, H., AMMAN, J., MÜLLER, W.: 7-Chlor und 7-Brom-Actinomycine. Tetrahedon Letters 3595-3597 (1966a).

BROCKMANN, H., LACKNER, H.: Totalsynthese von Actinomycin C_1 (D). Naturwissenschaften 51, 384-385 (1964a).

BROCKMANN, H., LACKNER, H.: Totalsynthese von Actinomycin C_2. Tetrahedon Letters 3517-3521 (1964b).

BROCKMANN, H., LACKNER, H.: Totalsynthese von Actinomycin C_3 über Bis-seco-actinomycin C_3. Chem. Ber. 1oo, 353-369 (1967).

BROCKMANN, H., MÜLLER, W., PETERSSEN-BORSTEL, H.: Actinomycin-Derivate aus Dihydro-Actinomycin. Tetrahedon Letters 3531-3535 (1966b).

BROCKMANN, H., SCHRAMM, W.: Synthese von Actinomycin-(thr-val-pro-sar-meval), dem Antipoden von Actinomycin C_1. Tetrahedon Letters 2331-2333 (1966).

BROCKMANN, H., SEELA, F.: Synthese von 1,8-Didesmethyl-Actinomycin C_1. Tetrahedon Letters 48o3-48o5 (1965).

CAVALIERI, L.F., NEMCHIN, R.G.: The mode of interaction of actinomycin D with deoxyribonucleic acid. Biochim. biophys. Acta (Amst.) 87, 641-652 (1964).

CAVALIERI, L.F., NEMCHIN, R.G.: The binding of actinomycin D and F to bacterial DNA. Biochim. biophys. Acta (Amst.) 166, 722-725 (1968).

CERAMI, A., REICH, E., WARD, D.C., GOLDBERG, I.H.: The interaction of actinomycin with DNA; requirement for the 2-amino group of purines. Proc. nat. Acad. Sci. (Wash.) 57, 1o36-1o42 (1967).

CHEN, D., SARID, S., KATACHALSKI, E.: Studies on the nature of messenger RNA in germinating wheat embryos. Proc. nat. Acad. Sci. (Wash.) 6o, 9o2-9o9 (1968).

CHENG, T.-Y., SUEOKA, N.: Polymer similar to polydeoxyadenylate-thymidylate in various tissues of a marine crab. Science 143, 1442-1443 (1964).

CLEFFMANN, G.: Bildung zusätzlicher DNS nach Blockierung der Zellteilung von Tetrahymena durch Actinomycin. Z. Zellforsch. 7o, 29o-297 (1966).

COLEMAN, G., ELLIOTT, W.H.: Extracellular ribonuclease formation in B. subtilis by actinomycin D. Nature 2o2, 1o83-1o85 (1964).

CONTI, F., De SANTIS, P.: Conformation of actinomycin-D. Nature 227, 1239-1241 (197o).

CROTHERS, D.M., SABOL, S.L., RATNER, D.I., MÜLLER, W.: Studies concerning the behaviour of actinomycin in solution. Biochemistry 7, 1817-1823 (1968).

DARNELL, J.E.: Ribonucleic acids from animal cells. Bact. Rev. 32, 262-29o (1968).

De SANTIS, P., RIZZO, R., UGHETTO, G.: Structure of actinomycin based on conformational studies. Nature New Biology 237, 94-95 (1972).

EAGLE, G.R., ROBINSON, D.S.: The ability of actinomycin D to increase the clearing-factor lipase activity of rat adipose tissue. Biochem. J. 93, 1oc-11c (1964).

ENGELS, W.: Zur Wikrung und Lokalisation von ^{3}H-Actinomycin D in Eifollikeln von Musca domestica nach in vivo-Applikation. Histochemie 19, 224-234 (1969).

FIELD, J.B., COSTA, F., BORYCZKA, A., SEKELY, L.I.: Experimental evaluation of the anticarcinogenic activity of a new antibiotic, actinomycin C. Antibiotics Annual, 1954-1955, New York 1955.

FIRTEL, R.A., BAXTER, L., LODISH, H.F.: Actinomycin D and the regulation of enzyme biosynthesis during development of Dictyostelium discoideum. J. molec. Biol. 79, 315-327 (1973).

FISCHER, H., MUNK, K.: Wirkung von Actinomycin D auf die DNA-Synthese in SV4o infizierten exponentiell wachsenden und stationären Zellen. Z. Krebsforsch. 74, 39o-395 (197o).

FOLEY, G.E., McCarthy, R.E., BINNS, V.M., SNELL, E.E., GUIRARD, B.M., KIDDER, G.W., DEWEY, V.C., THAYER, P.S.: A comparative study of the use of microorganisms in the screening of potential antitumour agents. Ann. N.Y. Acad. Sci. 76, 413-441 (1958).

FRANKLIN, R.M.: The inhibition of ribonucleic acid synthesis in mammalian cells by actinomycin D. Biochim. biophys. Acta (Amst.) 72, 555-565 (1963).

GARREN, L.D., HOWELL, R.R., TOMKINS, G.M., GROCCO, R.M.: A paradoxical effect of actinomycin D: The mechanism of regulation of enzyme synthesis by hydrocortisone. Proc. nat. Acad. Sci. (Wash.) 52, 1121-1129 (1964).

GELLERT, M., SMITH, C.E., NEVILLE, D., FELSENFELD, G.: Actinomycin binding to DNA; mechanism and specificity. J. molec. Biol. 11, 445-457 (1965).

GOLDBERG, I.H., FRIEDMAN, P.A.: Antibiotics and nucleic acids. Ann. Rev. Biochem. 4o, 775-81o (1971).

GOLDBERG, I.H., RABINOWITZ, M.R.: Actinomycin D inhibition of deoxyribonucleic cid-dependent synthesis of ribonucleic acid. Science 136, 315-316 (1962).

GOLDBERG, I.H., RABINOWITZ, M., REICH, E.: Basis of actinomycin action. I. DNA binding and inhibition of RNA-polymerase synthetic reactions by actinomycin. Proc. nat. Acad. Sci. (Wash.) 48, 2o94 to 21o1 (1962).

GOLDBERG, I.H., RABINOWITZ, M., REICH, E.: Basis of actinomycin action. II. Effect of actinomycin on the nucleoside triphosphate-inorganic pyrophosphate exchange. Proc. nat Acad. Sci. (Wash.) 49, 226-229 (1963).

GOLDSTEIN, M.N., SLOTNICK, I.J., HILLMANN, M.H., GALLAGHER, J.: Cyto-
chemical and biochemical studies on HeLa cells sensitive and resist-
ant to actinomycin D. Proc. Amer. Ass. Cancer Res. 3, 23-23 (1959).
GOMATOS, P.J., TAMM, I.: Animal and plant viruses with double-helical
RNA. Proc. nat. Acad. Sci. (Wash.) 5o, 878-885 (1963).
GURSKY, G.V.: Structure of DNA-actinomycin complex. Mol. Biol. SSSR
3, 749-757 (1969).
HACKMANN, C.: Experimentelle Untersuchungen über die Wirkung von Actinomycin (HBF 386) bei bösartigen Geschwülsten. Z. Krebsforsch.
58, 6o7-613 (1952).
HACKMANN, C.: HBF 386 (actinomycin C) ein cytostatisch wirksamer
Naturstoff. Strahlentherapie 9o, 296-3oo (1953).
HACKMANN, C.: Stoffwechselprodukte aus Mikroorganismen (Antibiotika)
als antineoplastische Wirkstoffe. Dtsch. med. Wschr. 8o, 812-818
(1955).
HAMELIN, R., LARSEN, C.J., TAVITIAN, A.: Effects of actinomycin D,
toyocamycin and cycloheximide on the synthesis of low-molecular-
weight nuclear RNAs in HeLa cells. Europ. J. Biochem. 35, 35o-356
(1973).
HAMILTON, L.D., FULLER, W., REICH, E.: X-ray diffraction and molecular model building studies of the interaction of actinomycin with
nucleic acids. Nature 198, 538-54o (1963).
HANDSCHACK, W., LINDIGKEIT, R.: Degradation of accumulated different
chloramphenicol-RNAs of Bacillus megaterium treatment with actinomycin D. Biochem. biophys. Res. Commun. 23, 793-798 (1966).
HARBERS, E., MÜLLER, W.: On the inhibition of RNA-synthesis by actinomycin. Biochem. biophys. Res. commun. 7, 1o7-11o (1962).
HARBERS, E., MÜLLER, W. BACKMANN, R.: Untersuchungen zum Wirkungsmechanismus der Actinomycine. II. Versuche mit ^{14}C-Actinomycin an
Ehrlich Ascitestumorzellen in vitro. Biochem. Z. 337, 224-231
(1963a).
HARBERS, E., BUJARD, H., MÜLLER, W.: Untersuchungen zum Wirkungsmechanismus der Actinomycine. IV. In vivo-Versuche mit ^{11}C-markiertem
C_1. III. International Congress of Chemotherapy 2, 995-1ooo. Stutt-
gart: Thieme 1963b.
HAREL, L., HAREL, J., BOER, A., IMBENOTTE, J., CARPENI, N.: Persistance
d'une synthèse de D-RNA dans le foie de rat traité par l'actinomycine. Biochim. biophys. Acta (Amst.) 87 212-218 (1964).
HARTMANN, G., COY, U.: Zum biologischen Wirkungsmechanismus des Actinomycine. Angew. Chem. 74, 5o1 (1962).
HARTMANN, G., COY, U., KNIESE, G.: Zum biologischen Wirkungsmechanismus der Actinomycine. Hoppe Seylers Z. physiol. Chem. 33o, 227-233
(1962).
HAYWOOD, A.M., HARRIS, J.M.: Actinomycin inhibition of MS 2 replication. J. molec. Biol. 18, 448-463 (1966).
HILL, R.B., Jr., SAUNDERS, E.H.: Actinomycin D and membrane polysome
interaction / abstrat rat liver RNA synthesis (Abstract). Amer. J.
Pathol. 55, 36A-37A (1969).
HO, P.P.K., WALTERS, C.P.: Influenza virus-induced ribonucleic acid
nucleotidyl-transferase and the effect of actinomycin D on its
formation. Biochemistry 5, 231-235 (1966).
HURWITZ, J., FURTH, J.J., MALAMY, M., ALEXANDER, M.: The role of deoxyribonucleic acid in ribonucleic acid synthesis. III. The inhibition of the enzymatic synthesis of ribonucleic acid and deoxyribonucleic acid by actinomycin D and proflavine. Proc. nat. Acad. Sci.
(Wash.) 48, 1222-123o (1962).
HYMAN, R.W., DAVIDSON, N.: Kinetics of the in vitro inhibition of
transcription by actinomycin. J. molec. Biol. 5o, 421-438 (197o).
JAIN, S.C., SOBELL, H.M.: Stereochemistry of actinomycin binding to
DNA. I. Refinement and further structural details of the actinomycin-deoxyguanosine crystalline complex. J. molec. Biol. 68, 1-2o
(1972).

KATZ, E.: Biogenesis of actinomycins. Ann. N.Y. Acad. Sci. 89, 3o4 to 322 (196o).

KATZ, E., WEISSBACH, H.: Biosynthesis of the actinomycin chromophor; enzymatic conversion of 4-methyl-3-hydroxy-anthranilic acid to actinocin. J. biol. Chem. 237, 882-886 (1962).

KAWAMATA, J., IMANISHI, M.: Interaction of actinomycin with DNA. Nature 187, 1112-1113 (196o).

KAWAMATA, J., IMANISHI, M.: Mechanism of action of actinomycin with special reference to its interaction with deoxyribonucleic acid. Biken's J. 4, 13-24 (1961).

KAWAMATA, J., OKUDAIRA, M., AKAMATSU, Y.: Autoradiographic studies on the intracellular distribution of ^{3}H-actinomycin S in TG cells. Biken's J. 8, 119-127 (1965).

KAY, J.E., COOPER, H.L.: Differential inhibition of 28S and 18S ribosomal RNA synthesis by actinomycin. Biochem. biophys. Res. Commun. 35, 526-53o (1969).

KELLY, T.J., SMITH, H.O.: A restriction enzyme from Hemophilus influenzae. II. Base sequence of the recognition site. J. molec. Biol. 51, 393-4o9 (197o).

KERSTEN, W.: Reaktion von Actinomycin mit DNS und RNS. Symposium über Krebsprobleme 196o, S. 146-15o. Berlin-Göttingen-Heidelberg: Springer 1961a.

KERSTEN, W.: Interaction of actinomycin C with constituents of nucleic acid. Biochim. biophys. Acta (Amst.) 47, 61o-611 (1961b).

KERSTEN, W., KERSTEN, H.: Zur Wirkungsweise von Actinomycinen. II. Bildung überschüssiger Desoxyribonukleinsäure in Bacillus subtilis. Hoppe Seylers Z. physiol. Chem. 327, 234-242 (1962a).

KERSTEN, W., KERSTEN, H.: Zur Wirkungsweise von Actinomycinen. III. Bindung von Actinomycin C an Nukleinsäuren und Nukleotide. Hoppe Seylers Z. physiol. Chem. 33o, 21-3o (1962b).

KERSTEN, W., KERSTEN, H., RAUEN, H.M.: Action of nucleic acids on the inhibition of growth by actinomycin of Neurospora crassa. Nature 187, 6o-61 (196o).

KERSTEN, W., KERSTEN, H., SZYBALSKI, W.: Physico-chemical properties of complexes between deoxyribonucleic acid and antibiotics which affect ribonucleic acid synthesis. Biochemistry 5, 236-244 (1966).

KIRK, J.M.: The mode of action of actinomycin D. Biochim. biophys. Acta (Amst.) 42, 167-169 (196o).

KORN, D.: Inhibition of bacteriophage T4 deoxyribonucleic acid maturation by actinomycin D: The accumulation of "replicating deoxyribonucleic acid". J. biol. Chem. 242, 16o-162 (1967).

KORN, D., PROTASS, J.J., LEIVE, L.: A novel effect of actinomycin D in preventing bacteriophage T4 maturation in E. coli. Biochem. biophys. Res. Commun. 19, 473-481 (1965).

KRUGH, T.R.: Association of actinomycin D and deoxyribonucleotides as a model for binding of the drug to DNA. Proc. nat. Acad. Sci. (Wash.) 69, 1911-1914 (1972).

KRUGH, T.R., NEELY, J.W.: Actinomycin D-mononucleotide interactions as studied by proton magnetic resonance. Biochemistry 12, 1775 to 1782 (1973a).

KRUGH, T.R., NEELY, J.W.: Actinomycin D-deoxydinucleotide interactions as a model for binding of the drug to deoxyribonucleic acid. Proton magnetic resonance results. Biochemistry 12, 4418-4425 (1973b).

LADO, P., SCHWENDIMANN, M.: The effect of actinomycin on RNA synthesis and phosphate uptake in isolated castor bean cotyledons. A tentative evaluation of the half-lives of mRNAs for some enzymes. Ital. J. Biochem. 18, 138-153 (1969).

LEIVE, L.: Actinomycin sensitivity in E. coli. Biochem. biophys. Res. Commun. 18, 13-17 (1965a).

LEIVE, L.: A nonspecific increase in permeability in E. coli produced by EDTA. Proc. nat. Acad. Sci. (Wash.) 53, 745-75o (1965b).

LEIVE, L.: RNA degradation and the assembly of ribosomes in actinomycin treated E. coli. J. molec. Biol. 13, 862-875 (1965c).

LERMAN, L.S.: Structural considerations in the interaction of DNA
and acridines. J. molec. Biol. 3, 18-3o (1961).

LEVY, H.B.: Effect of actinomycin D on HeLa cell nuclear RNA metabolism. Proc. Soc. exp. Biol. (N.Y.) 113, 886-889 (1963).

LIERSCH, M., HARTMANN, G.: Die Bindung von Proflavin und Actinomycin
an Desoxyribonucleinsäure. II. Die Bindung an denaturierte und ein-
strängige DNA, Apyrimidinsäure and Apurinsäure. Biochem. Z. 343,
16-28 (1965).

LOH, P.C., SOERGEL, M.: Growth characteristics of reovirus type 2:
actinomycin and the synthesis of viral RNA. Proc. nat. Acad. Sci.
(Wash.) 54, 857-863 (1965).

LUNT, M.R., SINSHEIMER, R.L.: Inhibition of ribonucleic acid bacteriophage growth by actinomycin D. J. molec. Biol. 18, 541-546 (1966).

MACH, B., TATUM, E.L.: Ribonucleic acid synthesis in protoplasts of
E. coli: Inhibition by actinomycin D. Science 139, 1o51-1o52
(1963).

MAITRA, U., NAKATA, Y., HURWITZ, J.: The role of deoxyribonucleic
acid in ribonucleic acid synthesis. XIV. A study of the initiation
of ribonucleic acid synthesis. J. biol. Chem. 242, 49o8-4918 (1967).

McCOY, E.E., EBADI, M.: The paradoxical effect of hydrocortisone and
actinomycin on the activity of rabbit leucocyte alkaline phosphatase. Biochem. biophys. Res. Commun. 26, 265-271 (1967).

McDONNELL, J., GARAPIN, A.C., LEVINSON, W.E., QUINTRELL, N., FANSHIER,
L., BISHOP, J.M.: DNA polymerases of Rous sarcoma virus: delinea-
tion of two reactions with actinomycin. Nature 228, 433-435 (197o).

MERITS, I.: Actinomycin inhibition of RNA synthesis in rat liver.
Biochem. biophys. Res. Commun. 1o, 254-259 (1963).

MERITS, I.: Actinomycin inhibition of soluble ribonucleic acid synthesis in rat liver. Biochim. biophys. Acta (Amst.) 1o8, 578-582
(1965).

MOOG, F.: Intestinal phosphatase activity: acceleration of increase
by puromycin and actinomycin. Science 144, 414-416 (1964).

MOOG, F.: Acceleration of the normal and corticoid-induced increase
of alkaline phosphatase in the duodenum of the nursling mouse by
actinomycin D, puromycin, colchicine and ethionine. Advan. Enzyme
Reg. 3, 221-236 (1965).

MOSES, V., SHARP, P.B.: Effect of actinomycin on the synthesis of
macromolecules in E. coli. Biochim. biophys. Acta (Amst.) 119,
2oo-2o3 (1966).

MÜLLER, W., CROTHERS, D.M.: Studies of the binding of actinomycin and
related compounds to DNA. J. molec. Biol. 35, 251-29o (1968).

MÜLLER, W., EMME, I.: Zum Verhalten der Actinomycine in wässrigen
Lösungen. Z. Naturforsch. 2ob, 835-841 (1965).

MÜLLER, W., SPATZ, H.C.: Über die Struktur des Actinomycin-Desoxygua-
nosin-Komplexes. Z. Naturforsch. 2ob, 842-853 (1965).

NAKATA, A., SEKIGUCHI, M., KAWAMATA, J.: Inhibition of multiplication
of bacteriophage by actinomycin. Nature 189, 246-247 (1961).

NITOWSKY, H., GELLER, S., CASPER, R.: Effects of actinomycin on induction of alkaline phosphate in heteroploid cell cultures. Fed.
Proc. 23, 556 (1964).

NITTA, K.: Studies on the effects of actinomycetes products on the
culture of human carcinoma cells (strain HeLa). II. The effect of
known antitumour antibiotics on HeLa cells. Jap. J. med. Sci.
1o, 287-296 (1957).

OETTEL, H., WILHELM, G.: Vergleichende Prüfungen von 14 cytostatisch
wirksamen Produkten an 7 Tiertumoren. Naunyn-Schmiedeberg's Arch.
Pharmak. exp. Path. 23o, 559-593 (1957).

O'MALLEY, B.W.: In vitro hormonal induction of a specific protein
(avidin) in chick oviduct. Biochemistry 6, 2546-2551 (1967).

PAPACONSTANTINOU, J., STEWART, J.A., KOEHN, P.V.: A localized stimu-
lation of lens protein synthesis by actinomycin D. Biochim. biophys.
Acta (Amst.) 114, 428-43o (1966).
PERRY, R.P.: The cellular sites of synthesis of ribosomal and 4S RNA.
Proc. nat. Acad. Sci. (Wash.) 48, 2179-2186 (1963).
POLLOCK, M.R.: The differential effect of actinomycin D on the bio-
synthesis of enzymes in Bacillus subtilis and Bacillus cereus.
Biochim. biophys. Acta (Amst.) 76, 8o-93 (1963).
PULLMAN, B.: On the complexes of actinomycin with purines and deoxy-
ribonucleic acid. Biochim. biophys. Acta (Amst.) 88, 44o-441 (1964).
REICH, E., CERAMI, A., WARD, D.C.: Actinomycin. In: Antibiotics, Vol. 1:
Mechanism of Action (D. GOTTLIEB, P.D. SHAW, Eds.), pp. 714-725.
Berlin-Heidelberg-New York: Springer 1967.
REICH, E., FRANKLIN, R.M., SHATKIN, A.J., TATUM, E.L.: Effect of
actinomycin D on cellular nucleic acid synthesis and virus produc-
tion. Science 134, 556-557 (1961).
REICH, E., GOLDBERG, I.H.: Actinomycin and nucleic acid function.
Progr. Nucl. Acid. Res. 3, 183-234 (1964).
RICHARDSON, J.P.: Enzymic synthesis of RNA from T7 DNA. J. molec.
Biol. 21, 115-127 (1966a).
RICHARDSON, J.P.: The binding of RNA polymerase to DNA. J. molec.
Biol. 21, 83-114 (1966b).
RINGERTZ, N.R., BOLUND, L.: Actinomycin binding capacity of deoxy-
ribonucleoprotein. Biochim. biophys. Acta (Amst.) 174, 147-154
(1969).
ROBERTS, W.K., NEWMAN, J.F.E.: Use of low concentrations of actino-
mycin D in the study of RNA synthesis in Ehrlich ascites cells.
J. molec. Biol. 2o, 63-73 (1966).
ROBINSON, H.J., WAKSMAN, S.A.: Studies on the toxicity of actinomycin.
J. Pharmacol. exp. Ther. 74, 25-32 (1942).
ROSEN, F., RAINA, P.N., MILHOLLAND, R.J., NICHOL, C.A.: Induction of
several adaptive enzymes by actinomycin D. Science 146, 661-663
(1964).
SALZMAN, N.P., SHATKIN, A.J., SEBRING, E.D.: The synthesis of a
DNA-like RNA in the cytoplasm of HeLa cells infected with vaccina
virus. J. molec. Biol. 8, 4o5-416 (1964).
SARMA, D.S.R., REID, I.M., SIDRANSKY, H.: The selective effect of
actinomycin D on free polyribosomes of mouse liver. Biochem. biophys. Res. Commun. 36, 582-588 (1969).
SAUER, G., MUNK, K.: Interference of actinomycin D with the replication of the DNA of herpes virus. II. Relationship between yield of
virus and time of actinomycin treatment. Biochim. biophys. Acta
(Amst.) 119, 341-346 (1966).
SAUER, G., ORTH, H.D., MUNK, K.: Interference of actinomycin D with
the replication of the herpes virus DNA. I. Difference in behaviour
of cellular and viral nucleic acid synthesis following treatment
with actinomycin D. Biochim. biophys. Acta (Amst.) 119, 331-34o
(1966).
SCARANO, E., De PETROCELLIS, B., AUGUSTI-TOCCO, G.: Studies on the control of enzyme synthesis during the early embryonic development
of the sea urchins. Biochim. biophys. Acta (Amst.) 87, 174-176
(1964).
SCHARA, R., MÜLLER, W.: Über die Wechselwirkung des Actinomycin C$_3$
mit Mono-, Di- und Oligonucleotiden. Europ. J. Biochem. 29, 21o
bis 216 (1972).
SCHMIDT-KASTNER, G.: Actinomycin E und Actinomycin F, zwei neue bio-
synthetische Actinomycingemische. Naturwissenschaften 43, 131-132
(1956).
SCHOLTISSEK, C.: Unphysiological breakdown of fast-labelled RNA by
actinomycin D in primary chick fibroblasts. Europ. J. Biochem. 28,
7o-73 (1972).

SEKIGUCHI, M., IIDA, S.: Mutants of E. coli permeable to actinomycin. Proc. nat. Acad. Sci. (Wash.) 58, 2315-232o (1967).

SHATKIN, A.J.: Actinomycin inhibition of ribonucleic acid synthesis and poliovirus infection of HeLa cells. Biochim. biophys. Acta (Amst.) 61, 31o-313 (1962).

SHATKIN, A.J.: Actinomycin and the differential synthesis of reovirus and L cell RNA. Biochem. biophys. Res. Commun. 19, 5o6-51o (1965).

SLOTNICK, I.J.: Disproportionate production of DNA in Bacillus subtilis inhibited by actinomycin D. Bact. Proc. 59, 13o-13o (1959).

SOBELL, H.M.: The stereochemistry of actinomycin binding to DNA and its implications in molecular biology. Progr. Nucl. Acid Res. molec. Biol. 13, 153-19o (1973).

SOBELL, H.M., JAIN, S.C.: Stereochemistry of actinomycin binding to DNA. II. Detailed molecular model of actinomycin-DNA complex and its implications. J. molec. Biol. 68, 21-34 (1972).

SOBELL, H.M., JAIN, S.C., SAKORE, T.D., NORDMAN, C.E.: Stereochemistry of actinomycin-DNA binding to DNA through formation of crystalline complex. Nature New Biology 231, 2oo-2o5 (1971a).

SOBELL, H.M., JAIN, S.C., SAKORE, T.D., PONTICELLO, G., NORDMAN, C.E.: Concerning the stereochemistry of actinomycin binding to DNA: An actinomycin-deoxyguanosine crystalline complex. Cold Spr. Harb. Symp. quant. Biol. 36, 263-27o (1971b).

SPIEGELMAN, S., BURNY, A., DAS, M.R., KEYDAR, J., SCHLOM, J., TRAVNICEK, M., WATSON, K.: DNA-directed DNA polymerase activity in oncogenic RNA viruses. Nature 227, 1o29-1o31 (197o).

STAEHELIN, T., WETTSTEIN, F.O., NOLL, H.: Breakdown of rat liver ergosomes in vivo after actinomycin inhibition of mRNA synthesis. Science 14o, 18o-183 (1963).

STRECKER, H.J., ELIASSON, E.E.: Ornithine δ-transaminase activity during the growth cycle of Chang's liver cells. J. biol. Chem. 241, 575o-5756 (1966).

SUEOKA, N.: Variation and heterogeneity of base composition of deoxyribonucleic acids: a compilation of old and new data. J. molec. Biol. 3, 31-4o (1961).

TAMAOKI, T., MUELLER, G.C.: Synthesis of nuclear and cytoplasmic RNA of HeLa cells and the effect of actinomycin D. Biochem. biophys. Res. Commun. 9, 451-454 (1962).

TANNENBERG, W.J.K., SCHWARTZ, R.S.: Modification of the immune response by actinomycin. In: Actinomycin (S.A. WAKSMAN, Ed.), pp. 163-179. New York-London-Sydney: Wiley 1968.

TEMIN, H.M.: The effects of actinomycin D on growth of Rous sarcoma virus in vitro. Virology 2o, 577-582 (1963).

TEMIN, H.M., MIZUTANI, S.: RNA-dependent DNA polymerase in virions of Rous sarcoma virus. Nature 226, 1211-1213 (197o).

THOMPSON, E.B., GRANNER, D.K., TOMKINS, G.M.: Superinduction of tyrosine aminotransferase by actinomycin D in rat hepatoma (HTC) cells. J. molec. Biol. 54, 159-175 (197o).

WAKSMAN, S.A., WOODRUFF, H.B.: Bacteriostatic and bactericidal substances produced by a soil Actinomyces. Proc. Soc. exp. Biol. (N.Y.) 45, 6o9-614 (194o).

WEISSBACH, H., KATZ, E.: Studies on the biosynthesis of actinomycin: Enzymic synthesis of the phenoxazone chromophore. J. biol. Chem. 236, PC16-PC18 (1961).

WEISSBACH, H., REDFIELD, B., BEAVEN, V., KATZ, E.: 4-Methyl-3-hydroxyanthranilic acid, an intermediate in actinomycin biosynthesis. Biochem. biophys. Res. Commun. 19, 524-53o (1965).

WELLS, R.D.: Actinomycin binding to DNA: Inability of a DNA-containing guanine to bind actinomycin D. Science 165, 75-76 (1969).

WELLS, R.D., LARSON, J.E.: Studies on the binding of actinomycin D to DNA and DNA model polymers. J. molec. Biol. 49, 319-342 (197o).

YAMAOKA, K., ZIFFER, H.: The optical properties of actinomycin D.
  II. Optical activity of the deoxyribonucleic acid complex. Biochemistry 7, 1oo1-1oo8 (1968).
ZIFFER, H., YAMAOKA, K., MAUGER, A.B.: Optical properties of actinomycin D. I. Influence of the lactone rings on its optical activity.
  Biochemistry 7, 996-1oo1 (1968).
ZIPPER, P., KRATKY, O., BUNEMANN, H., MÜLLER, W.: A small-angle X-ray
  scattering study on the interaction of actinomycin C_3 with deoxyribonucleic acid from calf thymus. FEBS Letters 25, 123-126 (1972).

B. Anthracyclines

AKASAKA, K., SAKODA, M., HIROMI, K.: Kinetic studies on acridine orange-DNA interaction by fluorescence stopped-flow method. Biochem.
  biophys. Res. Commun. 4o, 1239-1245 (197o).
AKTIPIS, S., KINDELIS, A.: Optical properties of the deoxyribonucleic
  acid-ethidium bromide complex. Effect of salt. Biochemistry 12,
  1213-1221 (1973).
ANGIULI, R., FORESTI, E., RIVA DI SANSEVERINO, L., ISAACS, N.W.,
  KENNARD, O., MOTHERWELL, W.D.S., WAMPLER, D.L., ARCAMONE, F.: Struc-
  ture of daunomycin; X-ray analysis of N-Br-acetyl-daunomycin sol-
  vate. Nature New Biology 234, 78-8o (1971).
ARCAMONE, F., FRANCESCHI, G., OREZZI, P., CASSINELLI, G., BARBIERI,
  W., MONDELLI, R.: Daunomycin. I. The structure of daunomycinone.
  J. Amer. chem. Soc. 86, 5334-5335 (1964a).
ARCAMONE, F., CASSINELLI, G., OREZZI, P., FRANCESCHI, G., MONDELLI,
  R.: Daunomycin. II. The structure and stereochemistry of daunosami-
  ne. J. Amer. chem. Soc. 86, 5335-5336 (1964b).
ARCAMONE, F., FRANCESCHI, G., PENCO, S., SELVA, A.: Adriamycin (14-
  hydroxy daunorubicin) a novel antitumour antibiotic. Tetrahedron
  Letters pp. 1oo7-1o1o (1969).
BARTHELEMY-CLAVEY, V., MAURIZOT, J.-C., SICARD, P.J.: Etude spectro-
  photométrique du complexe DNA-daunorubicine. Biochimie 55, 859-868
  (1973).
BAUER, W., VINOGRAD, J.: The interaction of closed circular DNA with
  intercalative dyes. I. The superhelix density of SV 4o DNA in the
  presence and absence of dye. J. molec. Biol. 33, 141-171 (1968).
BAUER, W., VINOGRAD, J.: The interaction of closed circular DNA with
  intercalative dyes. II. The free energy of superhelix formation,
  in SV 4o DNA. J. molec. Biol. 47, 419-435 (197o).
BAUER, W., VINOGRAD, J.: The use of intercalative dyes in the study
  of closed circular DNA. In: Progr. molec. subcell. Biol. (F.E. HAHN,
  Ed.), Vol. 2, pp. 181-215. Berlin-Heidelberg-New York: Springer 1971.
BHUYAN, B.K.: Phleomycin, xanthomycin, streptonigrin, nogalamycin and
  aurantin. In: D. GOTTLIEB, P.D. SHAW (Eds.): Antibiotics, Vol. I:
  Mechanism of Action, pp. 173-18o. Berlin-Heidelberg-New York:
  Springer 1967.
BHUYAN, B.K., DIETZ, A.D.: Fermentation, taxonomic and biological
  studies of nogalamycin. Antimicrob. Agents and Chemother. 1965,
  836-844 (1965).
BHUYAN, B.K., REUSSER, F.: Comparative biological activity of nogalamycin and its analogs. Cancer Res. 3o, 984-989 (197o).
BHUYAN, B.K., SMITH, C.G.: Differential interaction of nogalamycin
  with DNA of varying base composition. Proc. nat. Acad. Sci. (Wash.)
  54, 566-572 (1965).
BITTMANN, R.: Studies of the binding of ethidium bromide to transfer
  ribonucleic acid: absorption, fluorescence, ultracentrifugation and
  kinetic investigations. J. molec. Biol. 46, 251-268 (1969).

BLAKE, A., PEACOCKE, A.R.: The interaction of aminoacridines with
   nucleic acids. Biopolymers 6, 1225-1253 (1968).
BORISOVA, O.F., MINYAT, E.E.: Complexes of deoxynucleoprotein with
   acridine orange dye. Mol. Biol. SSSR 3, 758-767 (1969).
BRADLEY, D.F., LIFSON, S.: Statistical mechanical analysis of binding
   of acridines to DNA. In: B. PULLMAN (Ed.): Molecular Association
   in Biology, pp. 261-27o. New York-London: Academic Press 1968.
BRADLEY, D.F., WOLF, M.K.: Aggregation of dyes bound to polyanions.
   Proc. nat. Acad. Sci. (Wash.) 45, 944-952 (1959).
BRAZHNIKOVA, M.G., ZBARSKY, V.B., KUDINOVA, M.K., MURAVIEVA, L.I.,
   PONAMARENKO, V.I., POTAPOVA, N.P.: Carminomycin, a new antitumor
   anthracycline. Antibiotiki 18, 678-681 (1973a).
BRAZHNIKOVA, M.G., ZBARSKY, V.B., POTAPOVA, N.P., SHEINKER, Yu.N.,
   VLASOVA, T.F., ROZYNOV, B.V.: Structure of aglycone of carminomycin,
   an antitumour antibiotic. Antibiotiki 18, 1o59-1o63 (1973b).
BREMERSKOV, V., LINNEMANN, R.: Some effects of daunomycin on the nu-
   cleic acid synthesis in synchronized L-cells. Europ. J. Cancer 5,
   317-33o (1969).
BROCKMANN, H., BOLDT, P., NIEMEYER, J.: Beta-rhodomycinon and gamma-
   rhodomycinon. Chem. Ber. 96, 1356-1372 (1963).
CALENDI, E., Di MARCO, A., REGGIANI, M., SCARPINATO, B., VALENTINI,
   L.: On physico-chemical interactions between daunomycin and nucleic
   acids. Biochim. biophys. Acta (Amst.) 1o3, 25-49 (1965).
CANTOR, C.R., BEARDSLEY, K., NELSON, J., TAO, T., CHIN, K.W.: Studies
   on tRNA structure using covalently and noncovalently bound fluorescent dyes. In: Progr. molec. subcell. Biol. (F.E. HAHN, Ed.), Vol.2,
   pp. 297-315. Berlin-Heidelberg-New York: Springer 1971.
CHAN, E.W., BALL, J.K.: Interaction of DNA with three dimethyl derivatives of benz(c)acridines. Biochim. biophys. Acta (Amst.) 238,
   31-45 (1971).
CHURCHICH, J.E.: Fluorescence studies on soluble ribonucleic acid
   labelled with acriflavine. Biochim. biophys. Acta (Amst.) 75,
   274-276 (1963).
COHEN, G., EISENBERG, H.: Viscosity and sedimentation study of sonicated DNA-proflavine complexes. Biopolymers 8, 45-56 (1969).
COHEN, A., HARLEY, E.H., REES, K.R.: Antiviral effect of daunomycin.
   Nature 222, 36-38 (1969).
CRAWFORD, L.V., WARING, M.J.: Supercoiling of polyoma virus DNA
   measured by its interaction with ethidium bromide. J. molec. Biol.
   25, 23-3o (1967).
CROOK, L.E., REES, K.R., COHEN, A.: Effect of daunomycin on HeLa
   cell nucleic acid synthesis. Biochem. Pharmacol. 21, 281-286
   (1972).
DALGLEISH, D.G., FEY, G., KERSTEN, W.: Circular dichroism spectro-
   scopy of complexes of the antibiotics daunomycin, nogalamycin,
   chromomycin and mithramycin with DNA. Submitted for publication
   1973.
DANØ, K.: Development of resistance to adriamycin (NSC-123127) in
   Ehrlich ascites tumor in vivo. Cancer Chemother. Rep. Pt. 1 56,
   321-326 (1972).
Di MARCO, A.: Daunomycin and related antibiotics. In: D. GOTTLIEB,
   P.D. SHAW (Eds.): Antibiotics, Vol. I: Mechanism of Action, pp. 19o
   to 21o. Berlin-Heidelberg-New York: Springer 1967.
Di MARCO, A., SILVESTRINI, R., Di MARCO, S., DASDIA, T.: Inhibiting
   effect of the new cytotoxic antibiotic, daunomycin, on nucleic
   acids and mitotic activity of HeLa cells. J. Cell Biol. 27, 545
   to 55o (1965).
DUBOST, M., GANTER, P., MARAL, R., NINET, L., PINNERT, S., PREND'HOMME,
   J., WERNER, G.-H.: Rubidomycin: a new antibiotic with cytostatic
   properties. Cancer Chemother. Rep. 41, 35-36 (1964).

DOSKOČIL, J., FRIČ, I.: Complex formation of daunomycin with double-stranded RNA. FEBS Letters 37, 55-58 (1973).

DOURLENT, M., HÉLÈNE, C.: A quantitative analysis of proflavine binding to polyadenylic acid, polyuridylic acid, and transfer RNA. Europ. J. Biochem. 23, 86-95 (1971).

DUBOST, M., GANTER, P., MARAL, R., NINET, L., PINNERT, S., PREUD'HOMME, J., WERNER, G.-H.: Rubidomycin: a new antibiotic with cytostatic properties. Cancer Chemother. Rep. 41, 35-36 (1964).

ELLEM, K.A.O., RHODE III, S.L.: Selective inhibition of ribosomal RNA synthesis in HeLa cells by nogalamycin, a dA:dT binding antibiotic. Biochem. biophys. Acta (Amst.) 2o9, 415-424 (197o).

ETTLINGER, L., GÄUMANN, E., HÜTTER, R., KELLER-SCHIERLEIN, W., KRADOL-FER, F., NEIPP, L., PRELOG, V., REUSSER, P., ZÄHNER, H.: Stoffwechselprodukte von Actinomyceten. XVI. Cinerubine. Chem. Ber. 92, 1867-1879 (1959).

EVANS, I., LINSTEAD, D., RHODES, P.M., WILKIE, D.: Inhibition of RNA synthesis in mitochondria by daunomycin. Biochim. biophys. Acta 312, 323-326 (1973).

FELSTED, R.L., GEE, M., BACHUR, N.R.: Rat liver daunorubicin reductase an aldo-keto reductase. J. biol. Chem. 249, 3672-3679 (1974).

FEY, G.: Circulardichrographische Untersuchungen an DNA-Antibiotika-Komplexen. Diss. Nat.Fak. Erlangen-Nürnberg 1973.

FINKELSTEIN, T., WEINSTEIN, I.B.: Proflavine binding to transfer ribonucleic acid, synthetic ribonucleic acids, and deoxyribonucleic acid. J. Biol. Chem. 242, 3763-3768 (1967).

FULLER, W., WARING, M.J.: Molecular mode for the interaction of ethidium bromide with deoxyribonucleic acid. Ber. Bunsengesellsch. Phys. Chem. 68, 8o5-8o8 (1964).

GAUSE, G.F., SVESHNIKOVA, M.A., UKHOLINA, R.S., GAVRILINA, G.V., FILICHEVA, V.A., GLADKIKH, E.G.: Production of antitumor antibiotic carminomycin by Actinomadura carminate sp. nov. Antibiotiki 18, 675-678 (1973).

GRAY, G.D., CAMIENER, G.W., BHUYAN, B.K.: Nogalamycin effects in rat liver: Inhibition of tryptophan pyrrolase induction and nucleic acid biosynthesis. Cancer Res. 26, 2419-2424 (1966).

GREIN, A., SPALLA, C., Di MARCO, A., CANEVAZZI, G.: Descrizione e classificazione di un attinomicete (Streptomyces peucetius sp. nova) produttore di una sostanze ad attività antitumorale: la daunomycina. Giorn. Microbiol. 11, 1o9-118 (1963).

GROSJEAN, H., WÉRENNE, J., CHANTRENNE, H.: The binding of proflavine to transfer ribonucleic acid: Dependence on secondary structure. Biochim. biophys. Acta 166, 616-627 (1968).

HAHN, F.E. (Ed.): Complexes of biological active substances with nucleic acids and their modes of action. Progr. molec. subcell. Biol., Vol. 2. Berlin-Heidelberg-New York: Springer 1971.

HULTIN, T.: Effects of aminoacridines and related compounds on the conformation of rat liver ribosomes. Chem.-Biol. Interactions 2, 61-77 (197o).

INAGAKI, A., KAGEYAMA, M.: Interaction of antibiotics with deoxyribonucleic acid. II. DNA-cellulose chromatography of antibiotics and related compounds. J. Biochem. 68, 187-192 (197o).

IWAMOTO, R.H., LIM, P., BHACCA, N.S.: The structure of daunomycin. Tetrahedron Letters 3891-3894 (1968).

KELLER-SCHIERLEIN, W., RICHLE, W.: Metabolic products of microorganisms. LXXXVI. Structure of cinerubine A. Antimicr. Agents Chemother. 197o, 68-77 (197o).

KERSTEN, H., KERSTEN, W.: The failure of bacteria to control nucleic acid metabolism in the presence of sublethal doses of antibiotics. Proc. Congr. on Antibiotics, Prague 1964 (H. MILOS, G. ZDENEK, Eds.), pp. 645-65o. London: Butterworths 1966.

163

KERSTEN, W., KERSTEN, H.: Die Bindung von Daunomycin, Cinerubin und
  Chromomycin A$_3$ an Nucleinsäuren. Biochem. Z. 341, 174-183 (1965).
KERSTEN, W., KERSTEN, H.: Reaktionen verschiedener Antibiotika mit
  Nukleinsäuren und ihre Wirkung auf den Nukleinsäurestoffwechsel.
  Int. Symp. über Wirkungsmechanismen von Fungiziden und Antibiotika,
  S. 177-196. Berlin: Academie-Verlag 1967.
KERSTEN, W., KERSTEN, H.: Interaction of antibiotics with nucleic
  acids. In: B. PULLMAN (Ed.): Molecular Associations in Biology,
  pp. 289-298. New York-London: Academic Press 1968.
KERSTEN, W., KERSTEN, H., SZYBALSKI, W.: Physicochemical properties
  of complexes between deoxyribonucleic acid and antibiotics which
  affect ribonucleic synthesis (actinomycin, daunomycin, cinerubine,
  nogalamycin, chromomycin, mithramycin and olivomycin). Biochemistry
  5, 236-244 (1966).
KIM, J.H., GELBARD, A.S., DJORDJEVIC, B., KIM, S.H., PEREZ, A.G.:
  Action of daunomycin on the nucleic acid metabolism and viability
  of HeLa cells. Cancer Res. 28, 2437-2442 (1968).
KOSCHEL, K., HARTMANN, G., KERSTEN, W., KERSTEN, H.: Die Wirkung des
  Chromomycins und einiger Anthracyclinantibiotika auf die DNA-ab-
  hängige Nukleinsäure-Synthese. Biochem. Z. 344, 76-86 (1966).
LERMAN, L.: Structural considerations in the interaction of DNA and
  acridines. J. molec. Biol. 3, 18-3o (1961).
LERMAN, L.S.: The structure of the DNA-acridine complex. Proc. nat.
  Acad. Sci. (Wash.) 49, 94-1o2 (1963).
LERMAN, L.S.: Acridine mutagens and DNA structure. J. cell. comp.
  Physiol. 64, Suppl. 1-18 (1964).
LI, H.J., CROTHERS, D.M.: Relaxation studies of the proflavine-DNA
  complex: the kinetics of an intercalation reaction. J. molec. Biol.
  39, 461-477 (1969).
LÖBER, G.: Acridine, ihre physikochemische und biochemische Bedeutung.
  Z. Chem. 11, 92-1o2, 135-145 (1971).
LURQUIN, P., BUCHET-MAHIEU, J.: Biological activity of ethidium bro-
  mide-transfer RNA complexes. FEBS Letters 12, 244-248 (1971).
MITSCHER, L.A., McCRAE, W., ANDRES, W.W., LOWERY, J.A., BOHONOS, N.:
  Ruticulomycins, new anthracycline antibiotics. J. pharm. Sci. 53,
  1139-114o (1964).
NASS, M.M.K.: Differential effects of ethidium bromide on mitochron-
  drial and nuclear DNA synthesis in vivo in cultured mammalian cells.
  Exp. Cell Res. 72, 211-222 (1972).
NEWTON, B.A.: The mode of action of phenanthridines: the effect of
  ethidium bromide on cell division and nucleic acid synthesis. J.
  gen. Microbiol. 17, 718-73o (1957).
OSTERTAG, W., KERSTEN, W.: The action of proflavine and actinomycin
  D in causing chromatid breakage in human cells. Exp. Cell Res.
  39, 296-3o1 (1965).
PAOLETTI, J., Le PECQ, J.B.: Resonance energy transfer between ethi-
  dium bromide molecules bound to nucleic acids. Does intercalation
  wind or unwind the DNA helix?. J. molec. Biol. 59, 43-62 (1971).
PARISI, B., SOLLER, A.: Studies on the antiphage activity of daunomycin. Giorn. Microbiol. 12, 183-194 (1964).
PIGRAM, W.J., FULLER, W., HAMILTON, L.D.: Stereochemistry of inter-
  calation: interaction of daunomycin with DNA. Nature New Biology
  235, 17-19 (1972).
PULLMAN, B. (Ed.): Molecular association in biology. New York-London:
  Academic Press 1968.
REVEL, M., HIATT, H.H.: The stability of liver messenger RNA. Proc.
  nat. Acad. Sci. (Wash.) 51, 81o-818 (1964).
RINGERTZ, N.R., BOLUND, L., Dar ZYNKIEWICZ, Z.: AO binding of intra-
  cellular nucleic acids in fixed cells in relation to cell growth.
  Exp. Cell Res. 63, 233-238 (197o).

ROTH, D., MANJON, M.L.: Studies of a specific association between acriflavine and DNA in intact cells. Biopolymers 7, 695-7o6 (1969).

RUSCONI, A.: Different binding sites in DNA for actinomycin and daunomycin. Biochim. biophys. Acta (Amst.) 123, 627-63o (1966).

RUSCONI, A., Di MARCO, A.: Inhibition of nucleic acid synthesis by daunomycin and its relationship to the uptake of the drug into HeLa cells. Cancer Res. 29, 15o7-1511 (1969).

SAKODA, M., HIROMI, K., AKASAKA, K.: Kinetic studies on acridine orange-DNA interaction. A branched mechanism involving intercalation and outside "dimerization". J. Biochem. 71, 891-896 (1972).

SILVESTRINI, R., Di MARCO, A., DASDIA, T.: Interference of daunomycin with metabolic events of the cell cycle in synchronized culture of rat fibroblasts. Cancer Res. 3o, 966-973 (197o).

STEINERT, M., Van ASSEL, S., STEINERT, G.: Étude, par autoradiographie, des effects du bromure d'éthidium sur la synthèse des acides nucléiques de Crithidia luciliae. Exp. Cell. Res. 56, 69-74 (1969).

SUZUKI, T.: The structure of an antibiotic B-58941. Bull. chem. Soc. Jap. 43, 292 (197o).

TABACZYNSKI, M., SHELDRICK, P., SZYBALSKI, W.: Comparison of the mutagenic properties of the acridine dyes and of other "intercalating" agents, including ethidium bromide and the anthracycline antibiotics (cinerubin, daunomycin and nogalamycin). Microbial Genetics Bull. 23, 7-8 (1965).

TOMCHICK, R., MANDEL, H.G.: Biochemical effects of ethidium bromide in microorganisms. J. gen. Microbiol. 36, 225-236 (1964).

TRITTON, T.R., MOHR, S.C.: Relaxation kinetics of the binding of ethidium bromide to unfractionated yeast tRNA at low dye-phosphate ratio. Biochem. biophys. Res. Commun. 45, 124o-1249 (1971).

TRITTON, T.R., MOHR, S.C.: Kinetics of ethidium bromide binding as a probe of transfer ribonucleic acid structure. Biochemistry 12, 9o5-914 (1973).

TROUET, A., DEPREZ-de CAMPENEERE, D., de DUVE, C.: Chemotherapy through lysosomes with a DNA-daunorubicin complex. Nature New Biology 239, 11o-112 (1972).

ULLMAN, R.: Intrinsic viscosity of wormlike polymer chains. J. chem. Phys. 49, 5486-5497 (1968).

URBANKE, C., RÖMER, R., MAASS, G.: The binding of ethidium bromide to different conformations of tRNA. Unfolding of tertiary structure. Europ. J. Biochem. 33, 511-516 (1973).

WARING, M.J.: The effects of antimicrobial agents on ribonucleic acid polymerase. Molec. Pharmacol. 1; 1-13 (1965).

WARING, M.J.: Structural requirements for the binding of ethidium to nucleic acids. Biochim. biophys. Acta (Amst.) 114, 234-244 (1966).

WARING, M.J.: Intercalation into DNA. Naunyn-Schmiedebergs Arch. Pharmakol. 259, 91-97 (1968).

WARING, M.J.: Variation of the supercoils in closed circular DNA by binding ov antibiotics and drugs: evidence for molecular models involving intercalation. J. molec. Biol. 54, 247-279 (197o).

WARING, M.J.: Binding of drugs to supercoiled circular DNA: evidence for and against intercalation. In: F.E. HAHN(Ed.): Progr. molec. subcell. Biol., Vol. 2, pp. 216-231. Berlin-Heidelberg-New York: Springer 1971.

WILEY, P.F., MacKELLER, F.A., CARON, E.L., KELLY, R.B.: Isolation, characterization and degradation of nogalamycin. Tetrahedron Letters 663-668 (1968).

ZUNINO, F.: Studies on the mode of interaction of daunomycin with DNA. FEBS Letters 18, 249-253 (1971).

C. Chromomycin, Olivomycin and Mithramycin

BAKHAEVA, G.P., BERLIN, Y.A., BOLDYREVA, E.F., CHUPRUNOVA, O.A.,
    KOLOSOV, M.N., SOIFER, V.S., VASILJEVA, T.E., YARTSEVA, I.V.: The
    structure of aureolic acid (mithramycin). Tetrahedron Letters
    3595-3598 (1968).
BEHR, W., HARTMANN, G.: Spektralphotometrische Untersuchungen über
    die Wechselwirkungen zwischen Chromomycin A₃ und Nukleinsäuren.
    Biochem. Z. 343, 519-527 (1965).
BEHR, W., HONIKEL, K., HARTMANN, G.: Interaction of the RNA polymer-
    ase inhibitor chromomycin with DNA. Europ. J. Biochem. 9, 82-92
    (1969).
BERLIN, Y.A., ESIPOV, S.E., KOLOSOV, M.N., SHEMYAKIN, M.M.: The struc-
    ture of olivomycin. Tetrahedron Letters 1431-1436 (1966).
BRAZHNIKOVA, M.G., KRUGLYAK, E.B., BORISOVA, V.N., FEDOROVA, G.B.:
    A study of homogeneity of olivomycin. Antibiotiki 9, 141-146
    (1964a).
BRAZHNIKOVA, M.G., KRUGLYAK, E.B., MESENTSEV, A.S., FEDOROVA, G.B.:
    Products of acid hydrolysis of olivomycin. Antibiotiki 9, 552-553
    (1964b).
BRAZHNIKOVA, M.G., KRUGLYAK, E.B., KOVSHAROVA, I.N., KONSTANTINOVA,
    N.V., PROSHLYAKOVA, V.V.: Isolation, purification and study of
    certain physical-chemical properties of a new antibiotic olivomycin.
    Antibiotiki 7, 39-44 (1962).
FEY, G.: Circulardichrographische Untersuchungen an DNA-Antibiotika-
    Komplexen. Diss. Nat.Fak. Erlangen-Nürnberg 1973.
FEY, G., KERSTEN, H.: Circulardichrographie an Komplexen von Desoxy-
    ribonukleinsäuren mit Antibiotika. Hoppe-Seylers Z. physiol. Chem.
    351, 111 (1970).
GAUSE, G.F.: Chromomycin, olivomycin and mithramycin. In: D. GOTTLIEB,
    P.D. SHAW (Eds.): Antibiotics, Vol. 1: Mechanism of action, pp. 246
    to 258. Berlin-Heidelberg-New York: Springer 1967.
GAUSE, G.F., UCHOLINO, R.S., SVESHNIKOVA, M.A.: Olivomycin - a new
    antibiotic produced by Actinomyces olivoreticuli. Antibiotiki 7,
    34-38 (1962).
GAUSE, G.G., LOSHKAREVA, N.P.: Effect of olivomycin on the cells of
    Ehrlich ascites carcinoma. Vop. med. Khim. 11, 64-66 (1965).
GAUSE, G.G., LOSHKAREVA, N.P., DUDNIK, Y.V.: Mechanism of action of
    olivomycin. Antibiotiki 1o, 3o7-313 (1965).
HARTMANN, G., BEHR, W., BOCK, L., HONIKEL, K., LILL, H., LILL, U.,
    SIPPEL, A.: Zur Wirkung von Antibiotika auf die Nukleinsäuresynthe-
    se. Zbl. Bakt., I. Abt. Orig. 212, 224-232 (1970).
HARTMANN, ., GOLLER, H., KOSCHEL, K., KERSTEN, W., KERSTEN, H.:
    Hemmung der DNA-abhängigen RNA- und DNA-Synthese durch Antibiotika.
    Biochem. Z. 341, 126-128 (1964).
HAYASAKA, T., INOUE, Y.: Chromomycin A₃ studies in aqueous solutions.
    Spectrophotometric evidence for aggregation and interaction with
    herring sperm deoxyribonucleic acid. Biochemistry 8, 2342-2347
    (1969).
HONIKEL, K.O., SANTO, R.E.: A model for the in vitro inhibition of
    the DNA polymerase reaction with the base specific antibiotics
    Chromomycin A₃, Actinomycin-C₃ and Daunomycin. Biochim. biophys.
    Acta (Amst.) 269, 354-363 (1972).
KAMIYAMA, M., KAZIRO, Y.: Mechanism of action of chromomycin A₃.
    I. Inhibition of nucleic acid metabolism in B. subtilis cells.
    J. Biochem. (Tokyo) 59, 49-56 (1966).
KAZIRO, Y., KAMIYAMA, M.: Inhibition of RNA polymerase reaction by
    chromomycin A₃. Biochem. biophys. Res. Commun. 19, 433-437 (1965).
KERSTEN, W.: Interaction of carcinostatic antibiotics with nucleic
    acids. Gann Monograph 6, 65-71 (1968).

KERSTEN, W.: Inhibition of RNA synthesis by quinone antibiotics. In: Progr. molec. subcell. Biol., Vol. 2 (F.E. HAHN, Ed.), pp. 48 to 57. Berlin-Heidelberg-New York: Springer 1971.

KERSTEN, W., KERSTEN, H.: Die Bindung von Daunomycin, Cinerubin und Chromomycin A3 an Nukleinsäuren. Biochem. Z. 341, 174-183 (1965).

KERSTEN, W., KERSTEN, H.: Interaction of antibiotics with nucleic acids. In: B. PULLMAN (Ed.): Mol. Assoc. Biology, pp. 289-298. New York-London: Academic Press 1968.

KERSTEN, W., KERSTEN, H., STEINER, F.E., EMMERICH, B.: The effect of chromomycin and mithramycin on the synthesis of deoxyribonucleic acid and ribonucleic acids. Hoppe-Seylers Z. physiol. Chem. 348, 1415-1423 (1967).

KERSTEN, W., KERSTEN, H., SZYBALSKI, W.: Physicochemical properties of complexes between deoxyribonucleic acid and antibiotics which affect ribonucleic acid synthesis (actinomycin, daunomycin, cinerubine, nogalamycin, chromomycin, mithramycin and olivomycin). Biochemistry 5, 236-244 (1966).

KIDA, M., UJIHARA, M., OHMURA, E., KAZIWARA, K.: The effect of chromomycin A3 upon nucleic acid metabolism of B. subtilis SB-15. J. Biochem. 59, 353-362 (1966).

KOSCHEL, K., HARTMANN, G., KERSTEN, W., KERSTEN, H.: Die Wirkung des Chromomycins und einiger Anthracyclinantibiotika auf die DNA-abhängige Nukleinsäure-Synthese. Biochem. Z. 344, 76-86 (1966).

LAIKO, A.V.: The action of certain antitumour antibiotics on the synthesis of nucleic acids in cells of Staphylococci. Antibiotiki 7, 6o1-6o5 (1962).

MIYAMATO, M., MORITA, K., KAWAMATSU, Y., NOGUCHI, S., MARUMOTO, R., TANAKA, K., TATSUOKA, S., NAKANISHI, K., NAKADAIRA, Y., BHACCA, N.: Chromomycinone, the aglycone of chromomycin A3. Tetrahedron Letters 2355-2365 (1964a).

MIYAMOTO, M., MORITA, K., KAWAMATSU, Y., SASAI, M., NOHARA, A., TANAKA, K., TATSUOKA, S., NAKANISHI, K., NAKADAIRA, Y., BHACCA, N.: The structure of chromomycin A3. Tetrahedron Letters 2367-237o (1964b).

MIYAMOTO, M., KAWAMATSU, Y., SHINIHARA, M., NAKANISHI, K., NAKADAIRA, A., BHACCA, N.: The four chromoses from chromomycin A3. Tetrahedron Letters 2371-2377 (1964c).

MIYAMOTO, M., KAWAMATSU, Y., KAWASHIMA, K., SHINOHARA, M., NAKANISHI, K.: The full structures of three chromomycin A2, A3 and A4. Tetrahedron Letters 545-552 (1966).

MÜLLER, W., CROTHERS, D.M.: Studies of the binding of actinomycin and related compounds to DNA. J. molec. Biol. 35, 251-29o (1968).

NAYAK, R., SIRSI, M., PODDER, S.K.: Role of magnesium ion on the interaction between chromomycin A3 and deoxyribonucleic acid. FEBS Letters 3o, 157-162 (1973).

NORTHROP, G., TAYLOR, S.G., NORTHROP, R.L.: Biochemical effects of mithramycin on cultured cells. Cancer Res. 29, 1916-1919 (1969).

OSTERTAG, W., KERSTEN, W.: The action of proflavine and actinomycin D in causing chromatid breakage in human cells. Exp. Cell Res. 39, 296-3o1 (1965).

RAO, K.V., CULLEN, W.P., SOBIN, B.A.: A new antibiotic with antitumor properties. Antibiot. and Chemother. 12, 182-186 (1962).

RATAPONGS, C.: In vitro-Untersuchungen zur Wirkung von Chromomycin A3 auf den Nukleinsäurestoffwechsel von Ehrlich-Ascitestumorzellen. Naunyn-Schmiedebergs Arch. Pharmak. exp. Path. 262, 183-188 (1969).

SCHOLTISSEK, C., BECHT, H., MACPHERSON, I.: The effect of mithramycin on the multiplication of myxoviruses. J. gen. Virol. 8, 11-19 (197o).

SMITH, R., HENSON, D.: Inhibition of DNA virus replication with mithramycin. Fed. Proc. 24, 159 (1965).

TABACZYNSKI, M., SHELDRICK, P., SZYBALSKI, W.: Comparison of the mutagenic properties of the acridine dyes and of other "intercalating"
agents, including ethidium bromide and the anthracycline antibiotics
(cinerubin, daunomycin, and nogalamycin). Microbiol Genet. Bull.
23, 7-8 (1965).
TATSUOKA, S., NAKAZAWA, K., MIYAKE, A., KAZIWARA, K., ARAMAKI, Y.,
SHIBATA, M., TANABE, K., HAMADA, Y., HITOMI, H., MIYAMOTO, M.,
MIZUNO, K., WATANABE, J., ISHIDATE, M., YOKOTANI, H., USHIKAWA, I.:
Isolation, anticancer activity and pharmacology of a new antibiotic
chromomycin A_3. Gann 49 (suppl.), 23-24 (1958).
TATSUOKA, S., TANAKA, K., MIYAMOTO, M., MORITA, K., KAWAMATSU, Y.,
NAKANISHI, K., NAKADAIRA, Y., BHACCA, N.: The structure of chromo-
mycin A_3, a cancerostatic antibiotic. Proc. Jap. Acad. 4o, 236-24o
(1964).
WAKISAKA, G., UCHINO, H., NAKAMURA, T., SOTOBAYASHI, H., SHIRAKAWA,
S., ADACHI, A., SAKURAI, M.: Selective inhibition of biosynthesis
of ribonucleic acid in mammalian cells by chromomycin A_3. Nature
198, 385-386 (1963).
WARD, D., REICH, E., GOLDBERG, I.H.: Base specificity in the interaction of polynucleotides with antibiotic drugs. Science 149, 1259
to 1263 (1965).
WARING, M.: Variation of the supercoils in closed cirucular DNA by
binding of antibiotics and drugs: evidence for molecular models involving intercalation. J. molec. Biol. 54, 247-279 (197o).
YARBRO, J.W., KENNEDY, B.J., BARNUM, C.P.: Mithramycin inhibition of
ribonucleic acid synthesis. Cancer Res. 26, 36-39 (1966).
ZALMANZON, E.S., ZELENIN, A.V., KAFIANI, K.A., LOBAREVA, L.S., LYAPU-
NOVA, E.A., TIMOFEEVA, M.Y.: Effect of some antitumour antibiotics
on nucleic acid synthesis and virus reproduction in human amnion
cell cultures (strain FL). Antibiotiki 1o, 613-622 (1965).
ZALMANZON, E.S., ZELENIN, A.V., KAFIANI, K.A., LOBAREVA, L.S., LYAPU-
NOVA, E.A., TIMOFEEVA, M.Y.: On the mechanism of action of olivomycin. Vop. med. Khim 12, 52-62 (1966).

D. Kanchanomycin

BEERS, R.F., ARMILEI, G.: Heterogeneous binding of acridine orange
by polynucleotides. Nature 2o8, 466-468 (1965).
FRIEDMAN, P.A., JOEL, P.B., GOLDBERG, I.H.: Interaction of kanchanomycin with nucleic acids. I. Physical properties of the complex.
Biochemistry 8, 1535-1544 (1969a).
FRIEDMAN, P.A., LI, T.-K., GOLDBERG, I.H.: Interaction of kanchanomycin with nucleic acids. II. Optical rotatory dispersion and circular dichroism. Biochemistry 8, 1545-1553 (1969b).
JOEL, P.B., FRIEDMAN, P.A., GOLDBERG, I.H.: Interaction of kanchanomycin with nucleic acids. III. Contrasts in the mechanisms of inhibition of ribonucleic acid and deoxyribonucleic acid polymerase
ractions. Biochemistry 9, 4421-4427 (197o).
LIU, W.-C., CULLEN, W.P., RAO, K.V.: BA-18o265: A new cytotoxic antibiotic. Antimicrob. Agents Chemotherapy - 1962, 761-771 (1962).

E. Distamycin and Netropsin

ARCAMONE, F., PENCO, S., OREZZI, P., NICOLELLA, V., PIRELLI, A.:
Structure and synthesis of distamycin A. Nature 2o3, 1o64-1o65
(1964).

ARCAMONE, F., PENCO, S., Delle MONACHE, F.: Distamycin A. III. Synthesis of analogues with modifications in the side chains. Gazz. Chim. Ital. _99_, 62o-631 (1969).

ARCAMONE, F., NICOLELLA, V., PENCO, S., REDAELLI, S.: Distamycin A. IV. Synthesis of analogues with different number of residues of 1-methyl-4-aminopyrrole-2-carboxylic acid. Gazz. Chim. Ital. _99_, 632-64o (1969).

CASAZZA, A.M., GHIONE, M.: Therapeutic action of distamycin A on vaccinia virus infections in vivo. Chemotherapia (Basel) _9_, 8o-87 (1964/65).

CHANDRA, P., ZIMMER, C., THRUM, H.: Effect of distamycin A on the structure and template activity of DNA in RNA-polymerase system. FEBS Letters _7_, 9o-94 (197o).

CHANDRA, P., GÖTZ, A., WACKER, A., VERINI, M.A., CASAZZA, A.M., FIORETTI, A., ARCAMONE, F., GHIONE, M.: Some structural requirements for the antibiotic action of distamycins. FEBS Letters _16_, 249 to 252 (1971).

CHANDRA, P., GÖTZ, A., WACKER, A., ZUNINO, F., Di MARCO, A., VERINI, M.A., CASAZZA, A.M., FIORETTI, A., ARCAMONE, F., GHIONE, M.: Some structural requirements for the antibiotic action of distamycins, III. Possible interaction of formyl group of distamycin side chain with adenine. Hoppe Seylers Z. physiol. Chem. _353_, 393-398 (1972a).

CHANDRA, P., ZUNINO, F., GÖTZ, A., WACKER, A., GERICKE, D., Di MARCO, A., CASAZZA, A.M., GIULIANI, F.: Template specific inhibition of DNA polymerases from RNA tumour viruses by distamycin A and its structural analogues. FEBS Letters _21_, 154-158 (1972b).

Di MARCO, A., GAETANI, M., OREZZI, P., SCOTTI, T., ARCAMONE, F.: Experimental studies on distamycin A - a new antibiotic with cytotoxic activity. Cancer Chemother. Rep. _18_, 15-19 (1962).

Di MARCO, A., GHIONE, M., MIGLIACCI, A., MORVILLO, E., SANFILIPPO, A.: Studi sul meccanismo dell'azione antifagica dell'antibiotico distamicina. Giorn. Microbiol. _11_, 87-87 (1963a).

Di MARCO, A., GHIONE, M., SANFILIPPO, A., MORVILLO, E.: Selective inhibition of the multiplication of phage T1 in E. coli K12. Experientia (Basel) _19_, 134-136 (1963b).

ESTENSEN, R.D., KREY, A.K., HAHN, F.E.: Studies on a deoxyribonucleic acid-quinine complex. Molec. P'.armacol. _5_, 532-541 (1969).

FOURNEL, J., GANTER, P., KOENIG, F., de RATULD, J., WERNER, G.H.: Antiviral activity of distamycin A. Antimicr. Agents Chemother. - 1965, 599-6o4 (1965).

HAHN, F.E., O'BRIEN, R.L., CIAK, J., ALLISON, J.L., OLENICK, J.G.: Studies on modes of action of chloroquine, quinacrine and on chloroquine resistance. Milit. Med. _131_, 1o71-1o89 (1966).

KOTLER, M., BECKER, Y.: Rifampicin and distamycin A as inhibitiors of Rous sarcoma virus reverse transcriptase. Nature New Biology _234_, 212-214 (1971).

KREY, A.K., HAHN, F.E.: Studies on the complex of distamycin A with calf thymus DNA. FEBS Letters _1o_, 175-178 (197o).

PUSCHENDORF, B., GRUNICKE, H.: Effect of distamycin A on the template activity of DNA in a DNA polymerase system. FEBS Letters _4_,355 to 357 (1969).

PUSCHENDORF, B., PETERSEN, E., WOLF, H., WERCHAU, H., GRUNICKE, H.: Studies on the effect of distamycin A on the DNA dependent RNA polymerase system. Biochem. biophys. Res. Commun. _43_, 617-624 (1971).

REINERT, K.E.: Adenosine thymidine cluster-specific elongation and stiffening of DNA induced by the oligopeptide antibiotic netropsin. J. molec. Biol. _72_, 593-6o7 (1972).

VERINI, M.A., GIONE, M.: Activity of distamycin A on vaccinia virus infection of cell cultures. Chemotherapia (Basel) _9_, 145-16o (1964).

WERNER, G.H., GANTER, P., De RATULD, Y.: Studies on the antiviral activity of distamycin A. Chemotherapia _9_, 65-79 (1964).

ZILLIG, W., FUCHS, E., MILLETTE, R.: DNA-dependent RNA-polymerase.
In: Procedures in nucleic acid research (G.L. CANTONI, D.R. DAVIES,
Eds.), pp. 323-339. London: Harper and Row 1966.
ZIMMER, C., LUCK, G.: Optical rotatory dispersion properties of
nucleic acid complexes with the oligopeptide antibiotics distamycin
A and netropsin. FEBS Letters 1o, 339-342 (197o).
ZIMMER, C., LUCK, G.: Stability and dissociation of the DNA complexes
with distamycin A and netropsin in the presence of organic solvents,
urea and high salt concentration. Biochim. biophys. Acta (Amst.)
287, 376-385 (1972).
ZIMMER, C., PUSCHENDORF, B., GRUNICKE, H., CHANDRA, P., VENNER, H.:
Influence of netropsin and distamycin A on the secondary structure
and template activity of DNA. Europ. J. Biochem. 21, 269-278
(1971a).
ZIMMER, C., REINERT, E.-E., LUCK, G., WÄHNERT, U., LÖBER, G., THRUM,
H.: Interaction of the oligopeptide antibiotics netropsin and distamycin A with nucleic acids. J. molec. Biol. 58, 329-348 (1971b).
ZIMMER, C., LUCK, G., THRUM, H., PITRA, C.: Binding of analogues of
the antibiotics distamycin A and netropsin to native DNA. Europ.
J. Biochem. 26, 81-89 (1972).
ZUNINO, F., Di MARCO, A.: Studies on the interaction of distamycin A
and its derivatives with DNA. Biochem. Pharmacol. 21, 867-874
(1972).

F. Anthramycin

BATES, H.M., KUENZIG, W., WATSON, W.B.: Studies on the mechanism of
action of anthramycin-methyl-ether, a new antitumour antibiotic.
Cancer Res. 29, 2195-22o5 (1969)
GRUNBERG, E., PRINCE, H.N., TITSWORTH, E., BESKID, G., TENDLER, M.D.:
Chemotherapeutic properties of anthramycin. Chemotherapia (Basel)
11, 249-26o (1966).
HORWITZ, S.: Anthramycin. In. E. HAHN (Ed.): Progr. molec. subcell.
Biology, Vol. 2, pp. 4o-47. Berlin-Heidelberg-New York: Springer
1971.
HORWITZ, S.B., CHANG, S.C., GROLLMAN, A.P., BORKOVEC, A.B.: Chemo-
sterilant action of anthramycin: a proposed mechanism. Science
174, 159-161 (1971).
HORWITZ, S.B., GROLLMAN, A.P.: Interactions of small molecules with
nucleic acids. I. Mode of action of anthramycin. Antimicrobial
Agents and Chemotherapy - 1968, 21-24 (1968).
KOHN, K.W., BONO, V.H., jr., KANN, H.E., jr.: Anthramycin, a new type
of DNA-inhibiting antibiotic: reaction with DNA and effect on nucleic
acid synthesis in mouse leukemia cells. Biochim. biophys. Acta
(Amst.) 155, 121-129 (1968).
KOHN, K.W., SPEARS, C.L.: The reaction of anthramycin with DNA.
J. molec. Biol. 51, 551-572 (197o).
LEIMGRUBER, W., BATCHO, A.D., SCHENKER, F.: The structure of anthra-
mycin. J. Amer. chem. Soc. 87, 5793-5795 (1965a).
LEIMGRUBER, W., BATCHO, A.D., CZAJKOWSKI, R.C.: Total synthesis of
anthramycin. J. Amer. chem. Soc. 9o, 5641-5643 (1968).
LEIMGRUBER, W., STEFANOVIĆ, V., SCHENKER, F., KARR, A., BERGER, J.:
Isolation and characterization of anthramycin, a new antitumour
antibiotic. J. Amer. chem. Soc. 87, 5791-5793 (1965b).
STEFANOVIĆ, V.: Spectrophotometric studies of the interaction of
anthramycin with deoxyribonucleic acid. Biochem. Pharmacol. 17,
315-323 (1968).
TENDLER, M.D., KORMAN, S.: "Refuin" a non-cytotoxic carcinostatic
compound proliferated by a thermophilic actinomycete. Nature 199,
5o1-5o1 (1963).

III. Inhibitors of RNA Synthesis Interacting with RNA Polymerases

BLATTI, S.P., INGLES, C.J., LINDELL, T.J., MORRIS, P.W., WEAVER, R.F., WEINBERG, F., RÜTTER, W.J.: Structure and regulatory properties of eucaryotic RNA polymerase. Cold Spr. Harb. Symp. quant. Biol. 35, 649-657 (1970).
BURGESS, R.R.: RNA polymerase. Ann. Rev. Biochem. 4o, 711-74o (1971).
KEDINGER, D., NURET, P., CHAMBON, P.: Structural evidence for two α-amanitin sensitive RNA polymerases in calf thymus. FEBS Letters 15, 169-174 (1971).
POGO, A.O., LITTAU, V.C., ALLFREY, V.G., MIRSKY, A.E.: Modification of the ribonucleic acid synthesis in nuclei isolated from normal and regenerating liver: Some effects of salt and specific divalent cations. Proc. nat. Acad. Sci. (Wash.) 57, 743-75o (1967).
ROBERTS, J.W.: Termination factor of RNA synthesis. Nature 224, 1168-1174 (1969).
ROEDER, R.G., RUTTER, W.J.: Specific Nucleolar and Nucleoplasmic RNA Polymerases. Proc. nat. Acad. Sci. (Wash.) 65, 675-682 (1970).
TRAVERS, A.A., KAMEN, R.I., SCHLEIF, R.F.: Factor necessary for ribosomal RNA synthesis. Nature 228, 748-751 (1970).
WIDNELL, C.C., TATA, J.R.: Studies on the stimulation by ammonium sulphate of the DNA-dependent RNA polymerase of isolated rat liver nuclei. Biochim. biophys. Acta (Amst.) 123, 478-492 (1966).
ZUBAY, G., SCHWARTZ, D., BECKWITH, J.: Mechanism of activation of catabolite-sensitive genes: A positive control system. Proc. nat. Acad. Sci. (Wash.) 66, 1o4-11o (1970).

A. Rifamycin

BAUTZ, E., BAUTZ, F.: Initiation of RNA synthesis: The function of sigma in the binding of RNA-polymerase to promoter sites. Nature 226, 1219-1222 (1970a).
BAUTZ, E., BAUTZ, F.: Studies on the function of the RNA polymerase sigma factor in promoter selection. Cold. Spr. Harb. Symp. quant. Biol. 35, 227-232 (1970b).
BINDA, G., DOMENICHINI, E., GOTTARDI, A., ORLANDI, B., ORTELLI, E., PACINI, B., FOWST, G.: Rifampicin, a general review. Arzneimittel-Forsch. 21, 19o7-1977 (1971).
BLATTI, S.P., INGLES, C.J., LINDELL, T.J., MORRIS, P.W., WEAVER, R.F., WEINBERG, F., RUTTER, W.J.: Structure and regulatory properties of eucaryotic RNA polymerase. Cold Spr. Harb. Symp. quant. Biol. 35, 649-657 (1970).
BRUFANI, M., FEDELI, W., GIACOMELLO, G., VACIAGO, A.: The X-ray analysis of the structure of rifamycin B. Experentia (Basel) 2o, 339 to 342 (1964).
BUSIELLO, E., di GIROLAMO, A., di GIROLAMO, M., FISCHER-FANTUZZI, L., VESCO, C.: Multiple effects of rifamycin derivatives on animal-cell metabolism of macromolecules. Europ. J. Biochem. 35, 251-258 (1973).
CLEWELL, D.B., EVENCHIK, B., CRANSTON, J.W.: Direct inhibition of Col E1 plasmid DNA replication in E. coli by rifampicin. Nature New Biology 237, 29-31 (1972).
CLEWELL, D.B., EVENCHIK, B.G.: Effects of rifampicin, streptolydigin and actinomycin D on the replication of Col E1 plasmid DNA in Escherichia coli. J. molec. Biol. 75, 5o3-513 (1973).

DiMAURO, E., SNYDER, L., MARINO, P., LAMBERTI, A., COPPO, A., TOCCHINI-
  VALENTINI, G.P.: Rifampicin sensitivity of the components of DNA-
  dependent RNA polymerase. Nature 222, 533-537 (1969).
DOOLITTLE, W.F., PACE, N.R.: Synthesis of 5S ribosomal RNA in E. coli
  after rifamycin treatment. Nature 228, 125-129 (197o).
ENGELBERG, H.: Inhibition of RNA bacteriophage replication by rifam-
  picin. J. molec. Biol. 68, 541-546 (1972).
FRONTALI, L., TECCE, G.: Rifamycins. In: D. GOTTLIEB, P.D. SHAW (Eds.):
  Antibiotics, Vol. 1, Mechanism of Action, pp. 415-426. Berlin-
  Heidelberg-New York: Springer 1967.
GADALETA, M.N., GRECO, M., SACCONE, C.: The effect of rifampicin on
  mitochondrial RNA polymerase from rat liver. FEBS Letters 1o,
  54-56 (197o).
GALLO, R.C., YANG, S.S., SMITH, R.G., HERRERA, F., TING, R.C.,
  FUJIOKA, S.: Some observations on DNA polymerases of human normal
  and leukaemic cells. In: RIBBONS, D.W., WOESSNER, J.F., SCHULTZ, J.
  (Eds.): Nucleic Acid-Protein Interactions and Nucleic Acid Synthesis
  in Viral Infection, pp. 353-379. Amsterdam: North-Holland Publ. 1971.
GALLO, R.C., YANG, S.S., TING, R.C.: RNA-dependent DNA polymerase of
  human acute leukaemic cells. Nature 228, 927-929 (197o).
GERARD, G.F., GURGO, C., GRANDGENETT, D.P., GREEN, M.: Rifamycin
  derivatives: Specific inhibitors of nucleic acid polymerases.
  Biochem. biophys. Res. Commun. 53, 194-2o1 (1973).
GOLDBERG, I.H., FRIEDMAN, P.A.: Antibiotics and nucleic acids. Ann.
  Rev. Biochem. 4o, 775-81o (1971).
GRANT, W.D., POULTER, R.T.M.: Rifampicin-sensitive RNA and protein
  synthesis by isolated mitochondria of Physarum polycephalum. J.
  molec. Biol. 73, 439-454 (1973).
GURGO, C., GREEN, M.: Rifamycin derivatives strongly inhibiting RNA-
  DNA polymerase (reverse transcriptase) of murine sarcoma viruses.
  J. nat. Cancer Inst. 49, 61-79 (1972).
GURGO, C., RAY, R.K., THIRY, L., GREEN, M.: Inhibitors of the RNA
  and DNA dependent polymerase activities of RNA tumour viruses.
  Nature New Biology 229, 111-114 (1971).
HARFORD, N., SUEOKA, N.: Chromosomal location of antibiotic resistance
  markers in B. subtilis. J. molec. Biol. 51, 267-286 (197o).
HARTMANN, G., HONIKEL, K.O., KNÜSEL, F., NÜESCH, J.: The specific
  inhibition of the DNA-directed RNA synthesis by rifamycin. Biochim.
  biophys. Acta (Amst.) 145, 843-844 (1967).
HELLER, E., ARGAMAN, M., LEVY, H., GOLDBLUM, N.: Selective inhibition
  of vaccinia virus by the antibiotic rifampicin. Nature 222, 273 to
  274 (1969).
HINKLE, D.C., MANGEL, W.F., CHAMBERLIN, M.J.: Studies on the binding
  of E. coli RNA polymerase to DNA. IV. The effect of rifampicin on
  binding and on RNA chain initiation. J. molec. Biol. 7o, 2o9-22o
  (1972).
JOHNSTON, J.H., RICHMOND, M.H.: The increased rate of loss of penicillinase plasmids from Staph. aureus in the presence of rifampicin. J. gen. Microbiol. 6o, 137-139 (197o).
KATES, J.R., McAUSLAN, B.R.: Poxvirus DNA-dependent RNA polymerase.
  Proc. nat. Acad. Sci. (Wash.) 58, 134-141 (1967).
KERRICH-SANTO, R.E., HARTMANN, G.: On the formation of the rifampicin-
  resistant complex of RNA polymerase and DNA. Hoppe-Seylers Z. physiol. Chem. 353, 1535-1535 (1972).
KONINGS, R.N.H.: Personal communication.
KNÜSEL, F., BICKEL, H., KUMP, W.: A new group of rifamycin derivatives
  displaying activity against rifampicin resistant mutants of Staph.
  aureus. Experientia (Basel) 25, 12o7-12o8 (1969).
KRČMÉRY, V., JANOUSKOVA, J.: Effect of rifampicinon stability and
  transfer of R-factors. Z. allg. Mikrobiol. 11, 97-1o1 (1971).

LILL, U., HARTMANN, G.R.: Prevention of the binding of rifampicin to RNA polymerase. Hoppe Seylers Z. physiol. Chem. 353, 1544 (1972).

LILL, U.I., HARTMANN, G.R.: On the binding of rifampicin to the DNA-directed RNA polymerase from E. coli. Europ. J. Biochem. 38, 336-345 (1973).

LILL, H., LILL, U., SIPPEL, A., HARTMANN, G.: The inhibition of the RNA polymerase reaction by rifampicin. In: Proceedings of the 1st International Lepetit Colloquium on RNA Polymerase and Transcription (SILVESTRI, L., Ed.), pp. 55-64. London-Amsterdam: North Holland Publ. 1970.

MARGALITH, P., PAGANI, H.: Rifomycin. XIV. Production of rifomycin B. Appl. Microbiol. 9, 325-334 (1961).

MEIER, D., HOFSCHNEIDER, P.H.: Effect of rifampicin on the growth of RNA bacteriophage M12. FEBS Letters 25, 179-183 (1972).

MOSS, B., ROSENBLUM, E.N., KATZ, E., GRIMLEY, P.M.: Rifampicin: a specific inhibitor of vaccinia virus assembley. Nature 224, 1280 to 1284 (1960).

OPPOLZER, W., PRELOG, V., SENSI, P.: Konstitution des Rifamycins B und verwandter Rifamycine. Experientia (Basel) 20, 336-339 (1964).

PRELOG, V.: Über die Konstitution der Rifamycine. Chemotherapia (Basel) 7, 133-136 (1963a).

PRELOG, V.: Constitution of rifamycins. Pure Appl. Chem. 7, 551-564 (1963b).

RABUSSAY, D., ZILLIG, W.: A rifampicin resistant RNA polymerase from E. coli altered in the β-subunit. FEBS Letters 5, 104-106 (1969).

REID, P., SPEYER, J.: Rifampicin inhibition of ribonucleic acid and protein synthesis in normal and ethylenediaminetetraacetic acid-treated E. coli. J. Bact. 104, 376-389 (1970).

RIVA, S., FIETTA, A., SILVESTRI, L.G.: Mechanism of action of a rifamycin derivative (AF/013) which is active on the nucleic acid polymerases insensitive to rifampicin. Biochem. biophys. Res. Commun. 49, 1263-1271 (1972a).

RIVA, S., FIETTA, A.M., SILVESTRI, L.G., ROMERO, E.: Effect of rifampicin on expression of some episomal genes in E. coli. Nature New Biology 235, 78-80 (1972b).

RIVA, S., SILVESTRI, L.G.: Rifamycin: A General View. Ann. Rev. Microbiol. 26, 199-224 (1972).

ROMERO, E., RIVA, S., FIETTA, A.M., SILVESTRI, L.G.: Effect of R factors on rifampicin resistance in E. coli. Nature New Biology 234, 56-58 (1971).

RODRIGUEZ-LÓPEZ, M., MUÑOZ, M.L., VAZQUEZ, D.: The effect of rifamycin antibiotics on algae. FEBS Letters 9, 171-174 (1970).

SAROV, I., BECKER, Y.: RNA in the elementary bodies of trachoma agent. Nature 217, 849-852 (1968).

SENSI, P., GRECO, A.M., BALOTTA, R.: Rifomycins. I. Isolation and properties of rifomycin B and rifomycin complex. Antibiot. Ann. 1959-60, pp. 262-270.

SENTENAC, A., FROMAGEOT, P.: Effects of some RNA polymerase inhibitors on binding and chain initiation. In: T.H. BÜCHNER, H. SIES (Eds.): Mosbach Colloquium, pp. 64-75. Berlin-Heidelberg-New York: Springer 1969.

SHMERLING, Z.G.: The effect of rifamycin on RNA synthesis in the rat liver mitochondria. Biochem. biophys. Res. Commun. 37, 965-969 (1969).

SILVESTRI, L. (Ed.): RNA-polymerase and transcription. Proc. of the I. Int. Lepetit. Coll., pp. 1-339. Amsterdam-London: North Holland Publ. 1970.

SIPPEL, A., HARTMANN, G.: Mode of action of rifamycin on the RNA polymerase reaction. Biochim. biophys. Acta (Amst.) 157, 218-219 (1968).

SIPPEL, A., HARTMANN, G.: Rifampicin resistance of RNA polymerase in
    the binary complex with DNA. Europ. J. Biochem. 16, 152-157 (1970).
SMITH, R.G., WHANG-PENG, J., GALLO, R.C., LEVINE, P., TING, R.C.:
    Selective toxicity of rifamycin derivatives for leukaemic human
    leucocytes. Nature New Biology 236, 166-171 (1972).
STAUDENBAUER, W.L., HOFSCHNEIDER, P.H.: Replication of bacteriophage
    M 13: Inhibition of single-strand DNA synthesis by rifampicin.
    Proc. nat. Acad. Sci. (Wash.) 69, 1634-1637 (1972).
STRAAT, P.A., TS'O, P.O.: Ribonucleic acid polymerase from Micro-
    coccus luteus (Micrococcus lysodeikticus). IV. Effect of rifampi-
    cin and oligomers on the homopolymer-directed reaction. Biochemis-
    try 9, 926-931 (1970).
SUBAK-SHARPE, J.H., TIMBURY, M.C., WILLIAMS, J.F.: Rifampicin inhi-
    bits the growth of some mammalian viruses. Nature 222, 341-345
    (1969).
SURZYCKI, S.J.: Genetic functions of the chloroplast of Chlamydomonas
    reinhardi: Effect of rifampicin on chloroplast DNA-dependent RNA
    polymerase. Proc. nat. Acad. Sci. (Wash.) 63, 1327-1334 (1969).
TING, R.C., YANG, S.S., GALLO, P.C.: Reverse transcriptase, RNA
    tumour virus transformation and derivatives of rifamycin SV.
    Nature New Biology 236, 163-166 (1972).
TRAVERS, A.A., BURGESS, R.R.: Cyclic re-use of the RNA polymerase sigma
    factor. Nature 222, 537-540 (1969).
UMEZAWA, H., MIZUNO, S., YAMAZAKI, H., NITTA, K.: Inhibition of DNA-
    dependent RNA synthesis by rifamycins. J. Antibiot. (Tokyo) 21,
    234-236 (1968).
WEHRLI, W., KNÜSEL, F., SCHMID, K., STAEHELIN, M.: Interaction of
    rifamycin with bacterial RNA polymerase. Proc. nat. Acad. Sci.
    (Wash.) 61, 667-673 (1968b).
WEHRLI, W., NÜESCH, J., KNÜSEL, F., STAEHELIN, M.: Action of rifamy-
    cins on RNA polymerase. Biochim. biophys. Acta (Amst.) 157, 215 to
    217 (1968a).
WEHRLI, W., STAEHELIN, M.: The rifamycins - relation of chemical
    structure and action on RNA polymerase. Biochim. biophys. Acta
    (Amst.) 182, 24-29 (1969).
WEHRLI, W., STAEHELIN, M.: Interaction of rifamycin with RNA poly-
    merase. In: Proc. I. Int. Lepetit Colloq. on RNA polymerase and
    transcription, pp. 65-70. Amsterdam-London: North-Holland Publ.
    1970.
WEHRLI, W., STAEHELIN, M.: Actions of rifamycins. Bacteriol. Rev.
    35, 290-309 (1971).
WILHELM, J.M., OLEINICK, N.L., CORCORAN, J.W.: The inhibition of
    bacterial RNA synthesis by the rifamycin antibiotics. Biochim.
    biophys. Acta (Amst.) 166, 268-271 (1968).
WINSTEN, J.A., HUANG, P.C.: Ribosomal RNA synthesis in vitro: A
    protein-DNA complex from Bacillus subtilis active in initiation of
    transcription. Proc. nat. Acad. Sci. (Wash.) 69, 1387-1391 (1972).
WINTERSBERGER, E.: DNA-dependent RNA polymerase from mitochondria of a
    cytoplasmic "petite" mutant of yeast. Biochem. biophys. Res. Commun.
    40, 1179-1184 (1970).
WU, C.W., GOLDTHWAIT, D.A.: Studies of nucleotide binding to the RNA
    polymerase by a fluorescence technique. Biochemistry 8, 4450-4458
    (1969a).
WU, C.W., GOLDTHWAIT, D.A.: Studies on nucleotide binding to the RNA
    polymerase by equilibrium dialysis. Biochemistry 8, 4458-4464
    (1969b).
YANG, S.S., HERRERA, F.M., SMITH, R.G., REITZ, M.S., LANCINI, G.,
    TING, R.C., GALLO, R.C.: Rifamycin antibiotics: Inhibitors of
    Rauscher murine leukemia virus reverse transcriptase and of puri-
    fied DNA polymerase from human normal and leukemic lymphoblasts.
    J. nat. Cancer Inst. 49, 7-25 (1972).

ZILLIG, W., FUCHS, E., PALM, P., RABUSSAY, D., ZECHEL, K.: On the different subunits of DNA-dependent RNA polymerase from E. coli and their role in the complex function of the enzyme. In: Proc. I. Int. Lepetit Colloq. on RNA Polymerase and Transcription, pp. 151-157. Amsterdam-London: North-Holland Publ. 1970a.
ZILLIG, W., ZECHEL, K., RABUSSAY, D., SCHACHNER, M., SETHI, V.S., PALM, P., HEIL, A., SEIFERT, W.: On the role of different subunits of DNA-dependent RNA polymerase from E. coli in the transcription process. Cold Spr. Harb. Symp. quant. Biol. 35, 47-58 (1970b).

B. Streptovaricin

BORDEN, E.C., BROCKMAN, W.W., CARTER, W.A.: Selective inhibition by streptovaricin of splenomegaly induced by Rauscher leukaemia virus. Nature New Biology 232, 214-216 (1971).
BROCKMAN, W.W., CARTER, W.A., LI, L-H., REUSSER, F., NICHOL, F.R.: Streptovaricins inhibit RNA-dependent DNA polymerase present in an oncogenic RNA virus. Nature 230, 249-250 (1971).
CARTER, W.A., BROCKMAN, W.W., BORDEN, E.C.: Streptovaricins inhibit focus formation by MSV (MLV) complex. Nature New Biology 232, 212-214 (1971).
De BOER, C., MEULMAN, P.A., WNUK, R.J., PETERSON, D.H.: Geldanamycin, a new antibiotic. J. Antibiot. (Tokyo) 23, 442-447 (1970).
KAMIYA, K., SUGINO, T., WADA, Y., NISHIKAWA, M., KISHI, T.: The X-ray analysis of tolypomycinone tri-m-bromobenzoate. Experientia (Basel) 25, 901-903 (1969).
KISHI, T., ASAI, M., MUROI, M., HARADA, S., MIZUTA, E., TERAO, S., MIKI, T., MIZUNO, K.: Tolypomycin. I. Structure of tolypomycinone. Tetrahedron Letters 91-95 (1969a).
KISHI, T., HARADA, S., ASAI, M., MUROI, M., MIZUNO, K.: Tolypomycin. II. Structures of tolyposamide and tolypomycin Y. Tetrahedron Letters 97-100 (1969b).
MIZUNO, S., NITTA, K.: Effect of streptovaricin on RNA synthesis in phage T4-infected E. coli. Biochem. biophys. Res. Commun. 35, 127 to 130 (1969).
MIZUNO, S., YAMAZAKI, H., NITTA, K., UMEZAWA, H.: Inhibition of DNA-dependent RNA polymerase reaction of E. coli by an antimicrobial antibiotic, streptovaricin. Biochim. biophys. Acta (Amst.) 157, 322-332 (1968).
NITTA, S., MIZUNO, S., YAMAZAKI, H., UMEZAWA, H.: Streptovaricin- and rifampicin-resistance of RNA polymerase in a resistant clone of E. coli B. J. Antibiot. (Tokyo) 21, 521-522 (1968).
PIATIGORSKY, J., WHITELEY, A.H.: A change in permeability and uptake of [14C]uridine in response to fertilization in Strongylocentrotus purpuratus eggs. Biochim. biophys. Acta (Amst.) 108, 404-418 (1965).
RINEHART, K.L., jr., MAHESHWARI, M.L., ANTOSZ, F.J., MATHUR, H.H., SASAKI, K., SCHACHT, R.J.: Chemistry of the streptovaricins. VIII. Structures of streptovaricins, A, B, D, E, F and Gla. J. Amer. chem. Soc. 93, 6273-6274 (1971).
RHULAND, L.E., STERN, K.F., REAMES, H.R.: Streptovaricin. III. In vivo Studies in the tuberculous mouse. Amer. Rev. Tuberc. 75, 588 to 593 (1957).
SASAKI, K., KENNETH, L., RINEHART, K.L., jr., FREDERICK, J., ANTOSZ, F.J.: Chemistry of the streptovaricins. VI. Oxidation products from streptovaricin C. J. Antibiot. (Tokyo) 25, 68-70 (1972).
SIMINOFF, P., SMITH, R.M., SOKOLSKI, W.T., SAVAGE, G.M.: Streptovaricin. I. Discovery and biologic activity. Amer. Rev. Tuberc. 75, 576-583 (1957).

SOKOLSKI, W.T., EILERS, N.J., SIMINOFF, P.: Paper chromatography and assay of components in streptovarcin. Antibiot. Ann. (New York) 1957-1958, pp. 119-125.

TAN, K.B., McAUSLAN, B.R.: Inhibition of nucleoside incorporation into HeLa cells by streptovaricin. Biochem. biophys. Res. Commun. 42, 23o-236 (1971).

WANG, A.H.-J., PAUL, I.C., RINEHART, K.L., jr., ANTOSZ, F.J.: Chemistry of the streptovaricins. IX. X-ray crystallographic structure of a streptovaricin C derivative. J. Amer. chem. Soc. 93, 6275 to 6276 (1971).

WEHRLI, W., STAEHELIN, M.: The various actions of the rifamycins. Ciba-Geigy, Basel/Switzerland 1972.

WHITEFIELD, G.B., OLSON, E.C., HERR, R.R., FOX, J.A., BERGY, M.E., BOYACK, G.A.: Streptovaricin. II. Isolation and properties. Amer. Rev. Tuberc. 75, 584-587 (1957).

YURA, T., IGARASHI, K.: RNA polymerase mutants of E. coli. I. Mutants resistant to streptovaricins. Proc. nat. Acad. Sci. (Wash.) 61, 1313-1319 (1968).

YURA, T., IGARASHI, H., MASUKATA, K.: Temperature-sensitive RNA polymerase mutants of E. coli. In: Proc. I. Int. Lepetit Colloq. on RNA polymerase and transcription, pp. 71-89. Amsterdam-London: North-Holland Publ. 197o.

C. Streptolydigin

CASSANI, G., BURGESS, R.R., GOODMAN, H.M.: Streptolydigin inhibition of RNA polymerase. Cold Spr. Harb. Symp. quant. Biol. 35, 59-63 (197o).

CASSANI, G., BURGESS, R.R., GOODMAN, H.M., GOLD, L.: Inhibition of RNA polymerase by streptolydigin. Nature New Biology 23o, 197-2oo (1971).

De BOER, C., DIETZ, A., SILVER, S., SAVAGE, G.M.: Streptolydigin, a new antimicrobial antibiotic. I. Biological studies of streptolydigin. Antibiotics Ann. 1955-1956, 886-892.

EBLE, T.E., LARGE, C.M., DeVRIES, W.H., CRUM, G.F., SCHELL, J.W.: Streptolydigin, an antimicrobial antibiotic. II. Isolation and characterization. Antibiotics Ann. 1955-1956, 893-896.

RINEHART, K.L., jr., BECK, J.R., BORDERS, D.B., KINSTLE, T.H., KRAUSS, D.: Streptolydigin. III. Chromophore and structure. J. Amer. chem. Soc. 85, 4o38-4o39 (1963).

SCHLEIF, R.: Isolation and characterization of a streptolydigin resistant RNA polymerase. Nature 223, 1o68-1o69 (1969).

SIDDHIKOL, C., ERBSTOESZER, J.W., WEISBLUM, B.: Mode of action of streptolydigin. J. Bact. 99, 151-155 (1969).

Von der HELM, K., KRAKOW, J.S.: Inhibition of RNA polymerase by streptolydigin. Nature New Biology 235, 82-83 and 24o, 96 (1972).

ZILLIG, W., ZECHEL, K., RABUSSAY, D., SCHACHNER, M., SETHI, V.S., PALM, P., HEIL, A., SEIFERT, W.: On the rule of different subunits of DNA-dependent RNA polymerase from E. coli in the transcription process. Cold Spr. Harb. Symp. quant. Biol. 35, 47-58 (197o).

D. Amanitins

BEIDERBECK, R.: α-Amanitin hemmt die Tumorinduktion durch Agrobacterium tumefaciens. Z. Naturforsch. 27b, 1393-1394 (1972).

BLATTI, S.P., INGLES, C.J., LINDELL, T.J., MORRIS, P.W., WEAVER, R.F. WEINBERG, F., RUTTER, W.J.: Structure and regulatory properties of eucaryotic RNA polymerase. Cold Spr. Harb. Symp. 35, 649-657 (1970).

DEZELEE, S., SENTENAC, A., FROMAGEOT, P.: Study on yeast RNA polymerase. Effect of α-amanitin and rifampicin. FEBS Letters 7, 220-222 (1970).

FAULSTICH, H., WIELAND, T., JOCHUM, C.: Über die Inhaltstoffe des grünen Knollenblätterpilzes. Amanin und die Amanitine sind Sulfoxide. Liebigs Ann. Chem. 713, 186-195 (1968).

FIUME, L., La PLACA, M., PORTOLANI, M.: A study on the mechanism of the cytopathic action of α-amanitin. Sperimentale 116, 15-26 (1966).

FIUME, L., WIELAND, T.: Amanitins. Chemistry and action. FEBS Letters 8, 1-5 (1970).

JACOB, S.T., SAJDEL, E.M., MUNRO, H.N.: Specific action of α-amanitin on mammalian RNA polymerase protein. Nature 225, 60-62 (1970a).

JACOB, S.T., SAJDEL, E.M., MUNRO, H.N.: Different responses of soluble whole nuclear RNA polymerase and soluble nucleolar RNA polymerase to divalent cations and to inhibition by α-amanitin. Biochem. biophys. Res. Commun. 38, 765-770 (1970b).

JACOB, S.T., MUECKE, W., SAJDEL, E.M., MUNRO, H.N.: Evidence for extranucleolar control of RNA synthesis in the nucleolus. Biochem. biophys. Res. Commun. 40, 334-342 (1970c).

KEDINGER, C., GNIAZDOWSKI, M., MANDEL, J.L., jr., GISSINGER, F., CHAMBON, P.: α-Amanitin: a specific inhibitor of one of two DNA-dependent RNA polymerase activities from calf thymus. Biochem. biophys. Res. Commun. 38, 165-171 (1970).

KEDINGER, C., NURET, P., CHAMBON, P.: Structural evidence for two α-amanitin sensitive RNA polymerases in calf thymus. FEBS Letters 15, 169-174 (1971).

LINDELL, T.J., WEINBERG, F., MORRIS, P.W., ROEDER, R.G., RUTTER, W.J.: Specific inhibition of nuclear RNA polymerase II by α-amanitin. Science 170, 447-449 (1970).

MEIHLAC, M., KEDINGER, C., CHAMBON, P., FAULSTICH, H., GOVINDAN, M.V., WIELAND, T.: Amanitin binding to calf thymus RNA polymerase B. FEBS Letters 9, 258-260 (1970).

MONTANARO, N., NOVELLO, F., STIRPE, F.: Effect of α-amanitin on ribonucleic acid polymerase II of rat brain nuclei and on retention of avoidance conditioning. Biochem. J. 125, 1087-1090 (1971).

MONTECUCCOLI, G., NOVELLO, F., STIRPE, F.: Effect of α-amanitin poisoning on the synthesis of deoxyribonucleic acid and of protein in regenerating rat liver. Biochim. biophys. Acta 319, 199-208 (1973).

NIESSING, J., SCHNIEDERS, B., KUNZ, W., SEIFART, K.H., SEKERIS, C.E.: Inhibition of RNA synthesis by α-amanitin in vivo. Z. Naturforsch. 25b, 1119-1125 (1970).

NOVELLO, F., FIUME, L., STIRPE, F.: Inhibition by α-amanitin of ribonucleic acid polymerase solubilized from rat liver nuclei. Biochem. J. 116, 177-180 (1970).

POGO, A.O., LITTAU, V.C., ALLFREY, V.G., MIRSKY, A.E.: Modification of the ribonucleic acid synthesis in nuclei isolated from normal and regenerating liver. Proc. nat. Acad. Sci. (Wash.) 57, 743-750 (1967).

REEDER, R.H., ROEDER, R.G.: Ribosomal RNA synthesis in isolated nuclei. J. molec. Biol. 67, 433-441 (1972).

ROEDER, R.G., RUTTER, W.J.: Specific nucleolar and nucleoplasmic RNA polymerases. Proc. nat. Acad. Sci. (Wash.) 65, 675-682 (1970).

SEIFART, K.H., SEKERIS, C.E.: α-amanitin, a specific inhibitor of transcription by mammalian RNA polymerase. Z. Naturforsch. 24b, 1538-1544 (1969).

SHAAYA, E., CLEVER, U.: In vivo effects of α-amanitin on RNA synthesis
in Calliphora erythrocephala. Biochim. biophys. Acta (Amst.) 272,
373-381 (1972).
SHAAYA, E., SEKERIS, C.E.: Inhibitory effects of α-amanitin on RNA
synthesis and induction of DOPA-decarboxylase by β-ecdysone. FEBS
Letters 16, 333-336 (1971).
STIRPE, F., FIUME, L.: Studies on the pathogenesis of liver necrosis
by α-amanitin. Effect of α-amanitin on ribonucleic acid synthesis
and on ribonucleic acid polymerase in mouse liver nuclei. Biochem.
J. 1o5, 779-782 (1967).
TATA, J.R., HAMILTON, M.J., SHIELDS, D.: Effects of α-amanitin in vivo
on RNA polymerase and nuclear RNA synthesis. Nature New Biology
238, 161-164 (1972).
WIELAND, T.: Poisonous principles of mushrooms of the genus Amanita.
Science 159, 946-952 (1968).
WIELAND, T., GEBERT, U.: Über die Inhaltsstoffe des grünen Knollenblätterpilzes. XXX. Strukturen der Amanitine. Liebigs Ann. Chem.
7oo, 157-173 (1966).
WIELAND, T., WIELAND, O.: Chemistry and toxicology of the toxins of
Amanita phalloides. Pharmacol. Rev. 11, 87-1o7 (1959).
WIDNELL, C.C., TATA, J.R.: Studies on the stimulation by ammonium
sulphate of the DNA-dependent RNA polymerase of isolated rat-liver
nuclei. Biochim. biophys. Acta (Amst.) 123, 478-492 (1966).

IV. Inhibitors interferring at the precursor level or with regulatory
processes of nucleic acid synthesis

FILL, N.P., von MEYENBURG, K., FRIESEN, J.D.: Accumulation and turnover of guanosine tetraphosphate in Escherichia coli. J. molec.
Biol. 71, 769-783 (1972).

A. Nucleoside Antibiotics

ACS, G., REICH, E., MORI, M.: Biological and biochemical properties
of the analogue antibiotic tubercidin. Proc. nat. Acad. Sci. (Wash.
52, 493-5o1 (1964).
BLOCH, A., LEONARD, R.J., NICHOL, C.A.: On the mode of 7-deaza-adeno-
sine (tubercidin). Biochim. biophys. Acta (Amst.) 138, 1o-25
(1967).
BROWN, N.C.: 6-(p-Hydroxyphenylazo)-uracil: a selective inhibitor of
host DNA replication in phage-infected B. subtilis. Proc. nat.
Acad. Sci. (Wash.) 67, 1454-1461 (197o).
DARNELL, J.E., PHILIPSON, L., WALL, R., ADESNIK, M.: Polyadenylic
acid sequences: Role in conversion of nuclear RNA into messenger
RNA. Science 174, 5o7-51o (1971).
DAVIES, R.J.H.: Complex formation between polyadenylic acid and formycin B. J. molec. Biol. 73, 317-327 (1973).
GRAHN, B., LOVTRUP-REIN, H.: The effect of cordycepin on nuclear RNA
synthesis in nerve and glial cells. Acta physiol. scand. 82, 28-34
(1971).
OSSOWSKI, L., REICH, E.: Effects of nucleoside analogs on transcription of simian virus 4o. Virology 5o, 63o-639 (1972).

PENMAN, S., ROSBASH, M., PENMAN, M.: Messenger and heterogeneous
nuclear RNA in HeLa cells. Differential inhibition by cordycepin.
Proc. nat Acad. Sci. (Wash.) 67, 1878-1885 (197o).
PHILIPSON, L., WALL, R., GLICKMAN, G., DARNELL, J.E.: Addition of
polyadenylate sequences to virus-specific RNA during adenovirus
replication. Proc. nat. Acad. Sci. (Wash.) 68, 28o6-28o9 (1971).
PRUSINER, P., BRENNAN, T., SUNDARALINGAM, M.: Crystal structure and
molecular conformation of formycin monohydrates. Possible origin
of the anomalous cirucular dichroic spectra in formycin mono- and
polynucleotides. Biochemistry 12, 1196-12o2 (1973).
RAO, K.V., RENN, D.W.: BA-9o912; an antitumour substance. Antimicrobial Agent Chemother. 1963, 77-79 (1963).
RIZZO, A.J., KELLY, C., WEBB, T.E.: Effect of cordycepin on ribosome
formation and enzyme induction in rat liver. Canad. J. Biochem.
5o, 1o1o-1o15 (1972).
SHIGEURA, H.T., BOXER, G.E., MELONI, M.L., SAMPSON, S.D.: Structure-
activity relationship of some purine 3'-deoxyribonucleosides. Biochemistry 5, 994-1oo4 (1966).
SUHADOLNIK, R.J.: Nucleoside Antibiotics. New York-London-Sydney-
Toronto: Wiley-Interscience 197o.
SUHADOLNIK, R.J., UEMATSU, T., UEMATSU, H.: Toyocamycin: phosphorylation and incorporation into RNA and DNA and the biochemical properties of the triphosphate. Biochim. biophys. Acta (Amst.) 149,
41-49 (1967).
SVERAK, L., BONAR, R.A., LANGLOIS, A.J., BEARD, J.W.: Inhibition by
Toyocamycin of RNA synthesis in mammalian cells and in normal and
avian tumour virus-infected chick embryo cells. Biochim. biophys.
Acta (Amst.) 224, 441-45o (197o).
TRUMAN, J.T., FREDERIKSEN, S.: Effect of 3'-deoxyadenosine and 3'-
amino-3'-deoxyadenosine on the labelling of RNA sub-species in
Ehrlich ascites tumour cells. Biochim. biophys. Acta (Amst.) 182,
36-45 (1969).
TRUMAN, J.T., KLENOW, H.: Effect of 3'-amino-3'-deoxyadenosine on
nucleic acid synthesis in Ehrlich ascites tumour cells. Molec.
Pharmacol. 4, 77-86 (1968).
TAVITIAN, A., URETSKY, S.C., ACS, G.: Selective inhibition of ribo-
somal RNA synthesis in mammalian cells. Biochim. biophys. Acta
(Amst.) 157, 33-42 (1968).
TAVITIAN, A., URETSKY, S.C., ACS, G.: The effect of toyocamycin on
cellular RNA synthesis. Biochim. biophys. Acta (Amst.) 179, 5o-57
(1969).
WARD, D.C., CERAMI, A., REICH, E., ACS, G., ALTWERGER, L.: Biochemical studies of the nucleoside analogue, formycin. J. biol. Chem.
244, 3243-325o (1969).
WARD, D.C., REICH, E.: Conformational properties of polyformycin:
A polyribonucleotide with individual residues in the syn conformation. Proc. nat. Acad. Sci. (Wash.) 61, 1494-15o1 (1968).
WU, A.M., TING, R.C., PARAN, M., GALLO, R.C.: Cordycepin inhibits
induction of murine leukovirus production by 5-iodo-2'-deoxyuri-
dine. Proc. nat. Acad. Sci. (Wash.) 69, 382o-3824 (1972).

B. Mycophenolic Acids

ABRAHAM, E.: The effect of mycophenolic acid on the growth of Staphylococcus aureus in heart broth. Biochem. J. 39, 398-4o8 (1945).
ALSBERG, C.L., BLACK, O.F.: Bull. U.S. Bur. Pl. Ind. 27o, 7 (1913);
nach FRANKLIN, T.S., COOK, J.M.: Biochem. J. 113, 515 (1969).

CARTER, S.B.: U.K. Patent Application No. 43132/66 (1966).
CARTER, S.B., FRANKLIN, T.J., JONES, D.F., LEONARD, B.J., MILLS, S.C.,
   TURNER, R.W., TURNER, W.B.: Mycophenolic acid: an anti-cancer com-
   pound with unusual properties. Nature 223, 848-85o (1969).
FLOREY, H.W., GILLIVER, K., JENNINGS, M.A., SANDERS, A.G.: Mycophe-
   nolic acid, an antibiotic from Penicillium brevi compactum Dierckx.
   Lancet 25o, 46-49 (1946).
FRANKLIN, T.J., COOK, J.M.: The inhibition of nucleic acid synthesis
   by mycophenolic acid. Biochem. J. 113, 515-524 (1969).
GILLIVER, K.: The inhibitory action of antibiotics on plant patho-
   genic bacteria and fungi. Ann. Bot. 1o, 271-282 (1946).
GOSIO, B.: Richerche batteriologiche e chimiche sulle alterazione del
   mais; contributo all'etiologia della pellagra. Riv. d'ig. e san.
   publ., Roma 1896, VII, 825.
WILLIAMS, R.H., LIVELY, D.H., DeLONG, D.C., CLINE, J.C., SWEENEY,
   M.J., POORE, G.A., LARSEN, S.H.: Mycophenolic acid: antiviral and
   antitumour properties. J. Antibiot. (Tokyo) 21, 463-464 (1968).

C. Amino Acid Analogs

ANSFIELD, J.F.: Phase I study of azotomycin (NSC-56654). Cancer Chemother. Rep. 46, 37-4o (1965).
BENNETT, L.L., jr., SCHABEL, F.M., jr., SKIPPER, H.E.: Studies on the
   mode of action of azaserine. Arch. Biochem. 64, 423-436 (1956).
BROCKMAN, R.W., PITTILLO, R.F., SHADDIX, S., HILL, D.L.: Mode of action of azotomycin. Antimicrob. Agents Chemother. 1969, 56-62
   (1969).
CARTER, S.K.: Azotomycin (NSC-56654) clinical brochure. Cancer Chemo-
   ther. Rep. Part 3, 1, 2o7-217 (1968).
EHRLICH, J., COFFEY, G.L., FISHER, M.W., HILLEGAS, A.B., KOHBERGER,
   D.L., MACHAMER, H.E., RIGHTSEL, W.A., ROEGNER, F.R.: 6-Diazo-5-
   oxo-L-norleucine, a new tumour inhibitory substance. I. Biological
   studies. Antibiot. and Chemother. 6, 487-497 (1956).
FRANKLIN, T.J., SNOW, G.A.: Biochemistry of Antimicrobial Action.
   London: Chapman and Hall 1971.
GITTERMAN, C.P., DULANEY, E.L., KACZKA, G.A., HENDLIN, D., WOODRUFF,
   H.B.: The human tumour-egg host system. II. Discovery and proper-
   ties of a new antitumour agent, hadacidin. Proc. Soc. exp. Biol.
   (N.Y.) 1o9, 852-855 (1962).
HARRIS, J.J., TELLER, M.N., YAP-GUEVARA, E., WOOLEY, G.W.: Effects of
   hadacidin on human tumours grown in eggs and rats. Proc. Soc. exp.
   Biol. (N.Y.) 11o, 1-4 (1962).
HARTMAN, S.C., LEVENBERG, B., BUCHANAN, J.M.: Involvement of ATP,
   5-phosphoribosylpyrophosphate and L-azaserine in the enzymatic formation of glycinamide ribotide intermediates in inosinic acid biosynthesis. J. Amer. chem. Soc. 77, 5o1-5o3 (1955).
HENDERSON, J.F.: Feedback inhibition of purine biosynthesis in
   ascites tumour cells. J. biol. Chem. 237, 2631-2635 (1962).
LEVENBERG, G., MELNICK, I., BUCHANAN, J.M.: Biosynthesis of the purines. XV. The effect of aza-L-serine and 6-diazo-5-oxo-L-norleu-
   cine on inosinic acid biosynthesis de novo. J. biol. Chem. 225,
   163-176 (1957).
MOORE, E.C., HURLBERT, R.B.: Biosynthesis of RNA cytosine and RNA
   purines: differential inhibition by diazo-oxo-norleucine. Cancer
   Res. 21, 257-261 (1961).
RAO, K.V., BROOKS, S.C., KUGELMAN, M., ROMANO, A.A.: Diazomycins A,
   B and C, three antitumour substances. I. Isolation and characterization. Antibiot. Ann. 1959-196o, pp. 943-949.

SHIGEURA, H.T., GORDON, C.N.: Hadacidin, a new inhibitor of purine biosynthesis. J. biol. Chem. 237, 1932-1936 (1962).

TOMISEK, A.J., KELLY, J.H., SKIPPER, H.E.: Chromatographic studies of purine metabolism. I. The effect of azaserine on purine biosynthesis in E. coli using various ^{14}C-labelled precursors. Arch. Biochem. 64, 437-455 (1956).

WEISS, A.J., RAMIREZ, G., GRAGE, T., STRAWITZ, J., GOLDMAN, L., DOWNING, V.: Phase II study of azotomycin (NSC 56654). Cancer Chemother. Rep. 52, 611-614 (1968).

D. Quinone Antibiotics

CASHEL, M.: The control of ribonucleic acid synthesis in E. coli. IV. Relevance of unusual phosphorylated compounds from amino acid-starved stringent strains. J. biol. Chem. 244, 3133-3141 (1969).

CASHEL, M., KALBACHER, B.: The control of ribonucleic acid synthesis in E. coli. V. Characterization of a nucleotide associated with the stringent response. J. biol. Chem. 245, 23o9-2318 (197o).

CASHEL, M., LAZZARINI, R.A., KALBACHER, B.: An improved method for thin-layer chromatography of nucleotide mixtures containing ^{32}P-labelled orthophosphate. J. Chromatography 4o, 1o3-1o9 (1969).

CORBAZ, R., ETTLINGER, L., GÄUMANN, E., KALVODA, J., KELLER-SCHIERLEIN, W., KRADOLFER, F., MANUKIAN, B.K., NEIPP, L., PRELOG, V., REUSSER, P., ZÄHNER, H.: Stoffwechselprodukte von Actinomyceten: Granaticin. Helv. chim. Acta 4o, 1262-1269 (1957).

GRUNDMANN, E., JÜHLING, L., PÜTTER, J., SEIDEL, H.J.: Carcinostase durch heterocyclische Derivate des 2-Amino-1,4-Naphthochinons bei Transplantations-Tumoren. Z. Krebsforsch. 72, 185-196 (1969).

KELLER-SCHIERLEIN, W., BRUFANI, M., BARCZA, S.: Stoffwechselprodukte von Mikroorganismen (141). Die Struktur des Granaticins und des Granacitins B. 1. Teil: Spektroskopische Eigenschaften und chemischer Aufbau. Helv. chim. Acta 61, 1257-1268 (1968).

KERSTEN, H., KERSTEN, W.: Inhibitors acting on DNA and their use to study DNA replication and repair. In: T. BÜCHNER, E. SIES (Eds.): Inhibitors, tools in cell research, pp. 11-31. Berlin-Heidelberg-New York: Springer 197o.

KERSTEN, W.: Inhibition of RNA synthesis by quinone antibiotics. In: Progr. molec. subcell. Biol., Vol. 2, (F.E. HAHN, Ed.), pp. 48-57. Berlin-Heidelberg-New York: Springer 1971.

KERSTEN, W., KERSTEN, H., WANKE, H., OGILVIE, A.: Einfluß von Antibiotika auf Nukleinsäuren und den Nukleinsäurestoffwechsel. Zbl. Bakt. I. Abt. Orig. 212, 259-265 (1969).

KERSTEN, W., OGILVIE, A., WANKE, H.: Influence of granaticin and quinones on RNA- and protein metabolism. VI FEBS Meeting Madrid 1969, Abstract 676.

LAZZARINI, R.A., CASHEL, M., GALLANT, J.: On the regulation of guanosine tetraphosphate levels in stringent and relaxed strains of E. coli. J. biol. Chem. 246, 4381-4385 (1971).

OGILVIE, A.: Granaticin. Untersuchungen zum Wirkungsmechanismus. Diss. Münster 197o.

OGILVIE, A., KERSTEN, H., KERSTEN, W.: Hemmung der RNA-Synthese in Bakterien mit strenger Kontrolle (rel$^+$) durch Chinon-Antibiotika. Hoppe-Seylers Z. physiol. Chem. 353, 739-74o (1972).

OGILVIE, A., WIEBAUER, K., KERSTEN, H., KERSTEN, W.: Erhöhte Bildung von Guanosintetraphosphat bei verminderter Phosphorylierung von RNA-Vorstufen nach Behandlung von E. coli rel$^+$ mit Chinonantibiotika. Hoppe-Seylers Z. physiol. Chem. 353, 1555-1556 (1972).

OGILVIE, A., WIEBAUER, K., KERSTEN, W.: Accumulation of ppGpp and in-
   hibition of RNA synthesis in Bacteria with stringent control upon
   treatment with a synthetic quinone. IUB 9th Intern. Congr. of Bio-
   chemistry, Stockholm 1973, Handbook 315.
PÜTTER, J.: Über die Wirkung Hydroperoxyd-bildender Chinone auf Tumor-
   zellen. Hoppe-Seylers Z. physiol. Chem. 332, 1-16 (1963).
ROTH, R.H., REMERS, W.A., WEISS, M.J.: The mitomycin antibiotics.
   Synthetic studies. XIII. Indoloquinone analogs with variations at
   C-5. J. organ. Chem. 31, 1o12-1o15 (1966).
WANKE, H.: Wirkung natürlicher und synthetischer Chinone auf den
   Protein- und Nukleinsäurestoffwechsel. Diss. Münster 1971.
WANKE, H., KERSTEN, W., KERSTEN, H.: Polysomen in B. subtilis: Ein-
   fluß von Aminochinonen und Chinonantibiotica auf die Synthese und
   die Stabilität von mRNA. Hoppe-Seylers Z. physiol. Chem. 35o,
   1162-1163 (1969).
ZÄHNER, H.: Personal communication 1969.

Subject Index

Molecular Biology, Biochemistry and Biophysics

Editors: A. Kleinzeller, G.F. Springer, H.G. Wittmann

Distribution rights for U.K., Commonwealth, and the Traditional British Market (excluding Canada): Chapman & Halll, Ltd. London

Prices are subject to change without notice

Springer-Verlag
Berlin Heidelberg New York

■ Prospectus on request

Antibiotics

Volume 1
Mechanism of Action

Edited by D. Gottlieb and P.D. Shaw
With 197 figures. XII, 785 pages. 1967
Cloth DM 190,—; US $77.60
ISBN 3-540-03724-1

Volume 2
Biosynthesis

Edited by D. Gottlieb and P.D. Shaw
With 115 figures. XII, 466 pages. 1967
Cloth DM 120,—; US $49.00
ISBN 3-540-03725-X

The goal of the editors of these volumes
was to present as complete coverage as
possible of all information currently
available on the mechanism of action and
the biosynthesis of antibiotics. It includes
not only antibiotics on which there is
relatively definitive knowledge, but also
subjects on which the data is fragmentary
and less certain. The range of subjects
encompasses antibiotics of great medical
importance as well as those that are tools
for the loboratory investigator. Also included are antibiotics for which no utility
is known but which yet contribute to the
general state of knowledge in the field.
For almost all subjects, authors were
enlisted who had themselves worked with
the specific antibiotics. The authors were
encouraged to discuss the antibiotic as
part of a general biosynthetic process or
metabolic function, yet to keep the
information specific enough to convey in
detail the present state of knowledge for
the antibiotic. To overcome the difficulties
of comprehending specialized information
resulting from the fragmentation of
scientific knowledge, the authors sought
the antibiotic effects or syntheses to
normal metabolic processes. Exposition
of overall processes such as cell wall
synthesis, membrane permeability, and
protein synthesis are included. In the
case of Penicillin the mechanism of its
action in the actual infection has been
examined and the relation of drug action
to L forms of bacteria has been discussed.
The reveal information on different
antibiotics takes various roles: 1) as part
of the general scientific knowledge that
is vital to all future advances in this area,
2) as tools for the investigation of normal
metabolic behavior, 3) as the rational for
the use of antibiotics by the clinical and
general medical profession.

Volume 3
Mechanism of Action of
Antimicrobial and Antitumor Agents

Edited by J.W. Corcoran and F.E. Hahn
Assisted by J.F. Snell and K.L. Arora
With 197 figures. Approx. 750 pages. 1974
Cloth DM 188,—; US $76.70
ISBN 3-540-06653-5

This volume continues the collection of
review articles on the mechanism of action
of antibiotics, started in 1967 as
'Antibiotics 1-Mechanism of Action'
(edited by D. Gottlieb and P.D. Shaw).
The reader will find here the most up-
to-date information on the microbiology,
biochemistry, molecular biology and
pharmacology of drugs used in chemotherapy; both research and clinical
applications are covered.

Prices are subject to change without
notice

Springer-Verlag
Berlin Heidelberg New York

München Johannesburg London Madrid
New Delhi Paris Rio de Janeiro Sydney
Tokyo Utrecht Wien